the SALT LAKE VALLEY

"SALT LAKE VALLEY'S ENTERPRISES"
BY MARGARET GODFREY

PRODUCED IN COOPERATION WITH
THE SALT LAKE AREA CHAMBER OF COMMERCE

WINDSOR PUBLICATIONS, INC.
CHATSWORTH, CALIFORNIA

the SALT LAKE VALLEY

CHERYL A. SMITH

For Steve, my husband, who understands

Windsor Publications, Inc.—History Books Division
Managing Editor: Karen Story
Design Director: Alexander D'Anca

Staff for *The Salt Lake Valley: Crossroads of the West*
Manuscript Editor: Kevin Taylor
Photo Editor: Robin Mastrogeorge
Copy Editor: Bonnie Lund
Editor, Corporate Profiles: Brenda Berryhill
Production Editor, Corporate Profiles: Albert Polito
Proofreaders: Douglas P. Lathrop, Liz Reuben
Editorial Assistants: Phyllis Feldman-Schroeder, Kim Kievman, Michael Nugwynne,
    Kathy B. Peyser, Priscilla Solis, Theresa J. Solis
Publisher's Representatives, Corporate Profiles: Del Madsen, Gerry Harris, Beverly Cornell
Layout Artist, Corporate Profiles: E. Ifrah
Layout Artist, Editorial: Christina L. Rosepapa
Designer: Ellen Ifrah

Library of Congress Cataloging-in-Publication Data
Smith, Cheryl A., 1960-
The Salt Lake valley, crossroads of the West : a
contemporary portrait / by Cheryl A. Smith —1st ed.
    p. 208 cm. 23 x 31
    Bibliography: p. 203
    Includes index.
    ISBN 0-89781-306-5
    1.Salt Lake City (Utah)—Economic conditions. 2.Salt
Lake City (Utah)—Economic conditions—Pictorial works.
3.Great Salt Lake Valley (Utah)—Economic conditions.
4.Great Salt Lake Valley (Utah)—Economic conditions—Pictorial works. I.Title.
HC108.S57S57 1989
330.9792'42033—dc19   89-5489   CIP
ISBN: 0-89781-306-5

Windsor Publications, Inc.
Elliot Martin, Chairman of the Board
James L. Fish III, Chief Operating Officer
Michele Sylvestro, Vice President/Sales-Marketing

Frontispiece by Steve Greenwood

CONTENTS

Preface   9

Part One
"THIS IS THE PLACE"   11

Chapter One
A PIONEERING PLACE   12

The Salt Lake Valley's history, geographic location, and tradition of enterprise have influenced and enhanced the region's entrepreneurial spirit.

Chapter Two
WELCOME TO THE NEIGHBORHOOD   28

Salt Lake's hardworking residents, striking scenic beauty, and unparalleled quality of life attract people and businesses to the region.

Chapter Three
SALT LAKE MEANS BUSINESS   38

From mining and manufacturing to tourism, when it comes to business and industry, Salt Lake City is as modern as any city in the nation.

Chapter Four
BIONIC VALLEY   54

Salt Lake City is at the forefront of America's most significant biomedical discoveries and technological advancements, making Salt Lake's research community one of the most exciting and vibrant in the nation.

Chapter Five
CROSSROADS OF THE WEST   62

The efficient and accessible highways, railways, and airways of the Salt Lake Valley, together with its communication networks, connect the region to the world.

Chapter Six
HEALTHY, WEALTHY, AND WISE   74

A position of national leadership in education, along with innovative health care programs and advanced medical research, contribute to the Salt Lake Valley's exceptional quality of life.

Chapter Seven
THE GOOD LIFE   92

With an enthusiasm for culture, recreation, and fun, Salt Lake's residents have created a rich cosmopolitan environment with something for everyone.

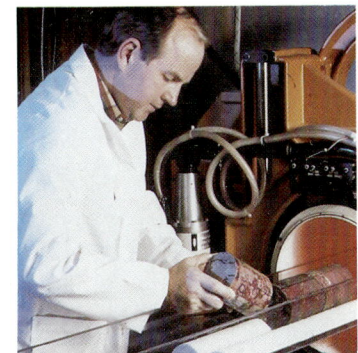

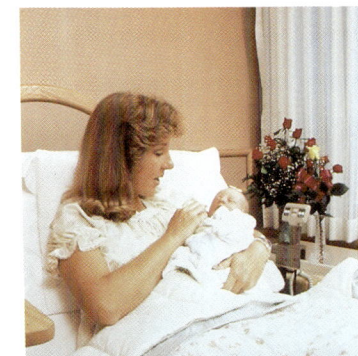

Chapter Eight
THE GREAT OUTDOORS 110

The unmatched natural beauty of the rugged peaks, alpine meadows, and desert expanses that surround the Salt Lake Valley enriches the lives of its residents, who strive to preserve it.

Part Two
SALT LAKE VALLEY'S ENTERPRISES 123

Chapter Nine
NETWORKS 124

The Salt Lake Valley's energy, communication, and transportation providers keep products, information, and power circulating within the area. Delta Air Lines, 126; Questar Corporation, 128; Intermountain Power Agency, 130; Utah Power & Light, 132; Deseret News, 133; Bonneville International Corporation, 134; KUTR-AM Radio, 136

Chapter Ten
MANUFACTURING AND MINING 138

Producing goods for individuals and industry, manufacturing and mining firms provide employment for many in the Salt Lake Valley. Mineral Mine, 140; Hercules Aerospace, 142; Morton International, 144; Thiokol Corporation, 145; Cytozyme, 146; Varian Eimac, 147; Geneva Steel of Utah, 148; Natter Manufacturing Company/Fairchild Industries, 151; Mark Steel Corporation, 152; Savage Industries, Inc., 154; Jetway Systems, 156; Huntsman Chemical Corporation, 158; Stabro Laboratories, Inc., 160; Deseret Medical, Inc., Becton Dickinson, 162; Eastman Christensen, 163

Chapter Eleven
BUSINESS AND PROFESSIONS 164

The Salt Lake Valley's business and professional community brings a wealth of service, ability, insight, and development into the area. Salt Lake Area Chamber of Commerce, 166; Utah Economic Development Corporation, 167; Arthur Andersen & Co., 168; Johnson & Higgins, 169; Snow, Christensen & Martineau, 170; IBM, 172; 3M Health Information Systems, 174; Van Cott, Bagley, Cornwall & McCarthy, 176; Jones, Waldo, Holbrook & McDonough, P.C., 177; Murdock Travel Management, 178; Peat Marwick Main & Co., 180; Babcock Pace + Associates, Architects, 181; City Centre/A Price-Prowswood Development, 182

Chapter Twelve
QUALITY OF LIFE 184

Medical and religious institutions contribute to the quality of life of Salt Lake Valley residents. Church of Jesus Christ of Latter-day Saints, 186; St. Mark's Hospital, 190

Chapter Thirteen
THE MARKETPLACE 192

The Salt Lake Valley's retail establishments, service industries, and products are enjoyed by residents and visitors alike. Solitude Ski Resort, 194; Crossroads Plaza Association, 195; The Snelgrove Ice Cream Company, 196; Orleans Inn Hotel, 198

Bibliography 202
Index 204

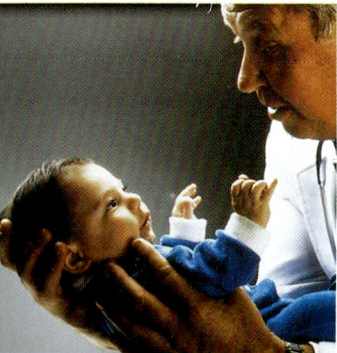

> "To maintain one's self on this Earth is not a hardship but a pastime, if we will live simply and wisely; as the pursuits of the simpler nations are still the sports of the more artificial."
>
> Henry D. Thoreau

PREFACE

I have titled this book *The Salt Lake Valley: Crossroads of the West* because Salt Lake is more than a city. It represents the promise of an entire region: the environment, the people, the heritage, the economics. My intention is that it serve as an introduction to some and a reminder to others.

In the beginning of this project, known simply as "the book" to family and friends, I couldn't imagine gathering enough information to fill a volume on one city. I soon realized, however, that I'd have a tougher time figuring out what facts, anecdotes, and descriptions to edit out without leaving a gaping hole in the end. I hope this is an informative profile, reference, and guide, but if, as Twain said, between the lines you find a story of your own, I guess I can live with that.

Not being an economist or a fortune teller with all the answers, I hope that this book is an honest portrait of the community—its people, heritage, and spectacular natural beauty.

Thanks to Paula and Fred for the recommendation, and to Kevin for the input. Thanks, too, to my family for saying "you can."

Salt Lake City seems to sparkle in the evening light. Photo by Audrey Gibson

Liberty Park provides a scenic and tranquil oasis in the heart of Salt Lake City. Photo by Mark Gibson

PART ONE

"THIS IS THE PLACE"

CHAPTER ONE

A PIONEERING PLACE

Salt Lake City wasn't settled by accident. Unlike many cities founded because of rich, fertile land, or established as convenient ports, Salt Lake City was originally settled as a religious refuge, where members of the fledgling Church of Jesus Christ of Latter-day Saints could live, work, and worship free from prejudice and persecution. Over the years Salt Lake City has grown from a reclusive community into a thriving center of business, transportation, health care, medical industries, education, and the arts.

The Great Salt Lake Valley, with its inland sea, towering mountains, and open spaces, has intrigued and attracted the adventurous, the curious, and the devout to its frontier for centuries. Some folks who passed through here a century ago are now famous legends, others are unsung heroes. Yet without either breed, Salt Lake City would not be quite the same today. Salt Lake City is a host for people from around the world; it is a city of people with bright ideas and innovative ways of doing things.

We strolled about everywhere through the broad, straight, level streets, and enjoyed . . . a grand general air of neatness, repair, thrift and comfort, around and about and over the whole. And everywhere were workshops, factories, and all manner of industries; and intent faces and busy hands were to be seen wherever one looked; and in one's ears was the ceaseless clink of hammers, the buzz of trade and the contented hum of drums and fly-wheels . . . Salt Lake was healthy—an extremely healthy city.

Mark Twain made those observations during his visit to Salt Lake City in the 1860s, and his remarks still hold true today. Of course, the city by the inland sea has grown up in this century, but some things have not changed.

Twain, like visitors to Salt Lake City today, was at once impressed by the contrasting geographic features of this great land:

At four in the afternoon we arrived on the summit of Big Mountain, fifteen miles from Salt Lake City, when all the world was glorified with the setting sun, and the most stupendous panorama of mountain peaks yet encountered burst on our sight.

The city lies at the edge of a level plain as broad as the state of Connecticut, and crouches close down to the ground under a curving wall of mighty mountains whose heads are hidden in the clouds, and whose shoulders bear relics of the snows of winter all the summer long . . . And on hot days in late spring and early autumn the citizens could quit fanning and growling and go out and cool off by looking at the luxury of a glorious snowstorm going on in the mountains.

When Twain traveled through Salt Lake at the age of 17 with his brother, 15,000 people lived here. The Salt Lake Temple and the University of Deseret (the present-day University of Utah) were already established, the *Deseret News* was being printed, and the city's wide, straight streets bordering oversized city blocks were bustling with activity. But the scene had not always been this civilized.

Just 20 years earlier this was a barren prairie of salt and sage—once the floor of ancient Lake Bonneville—yet it was also the promised land for thousands of members of the young

The developing Salt Lake Valley was rendered in this striking 1858 illustration, looking west toward the Great Salt Lake. Courtesy, Utah State Historical Society

Church of Jesus Christ of Latter-day Saints. Social and political pressures, prejudices, violence, and threats toward the rapidly growing population of Mormons forced thousands of the Saints from Nauvoo, Illinois. On February 4, 1846, led by Brigham Young, they began their harried flight westward, leaving behind their fertile farms, homes, prosperous businesses, furniture, and many personal belongings. They were seeking a secluded place where they could finally live and practice their faith in peace.

Standard supplies for a family of five included one solid wagon and three yoke of oxen, 1,000 pounds of flour, 22 pounds of sugar, two cows, and two beef cattle. Additional supplies consisted of a tent and bedding, one rifle and ammunition, cooking utensils, seeds for planting, and tools for farming.

Historians accompanying the expedition, and several pioneers who kept personal journals, documented the group's survival through winter's bitter cold and the endless spring rainstorms that mired the trails and made the rivers impassable. Disease, fever, chills, starvation, and exposure claimed hundreds of lives.

Historians estimate that in July 1846, 15,000 Mormons were camped or traveling

along the westward trails with 3,000 wagons and 30,000 horses, cattle, mules, and sheep. Until the famous California Gold Rush of 1849, this was the nation's largest human migration westward. The refugees built permanent settlements on the prairie at Garden Grove, Mount Pisgah, and Winter Quarters for those faithful who followed behind.

With the outbreak of the Mexican War, President Polk called for 500 able men from the thousands of migrating Mormons, leaving, for the most part, women, old men, and children to continue the trek. Ironically, the followers of Joseph Smith were now being asked to aid the very country from whose borders they were running to escape persecution. They already had been driven from three states where they had been persecuted for their beliefs and way of life, yet Mormon leader and Church president Brigham Young remained loyal to the flag. "You shall have your battalion," he said.

Through hardship and turmoil, the group attempted to keep their spirits high, their moods uplifted. Relaxing, singing, and dancing around the campfires at night seemed to keep up the travelers' morale.

In June 1847 they arrived at the Rocky Mountain Divide and were instantly cheered by the sight of waters flowing westward. They knew they were nearing their destination. Inspired by this vision, the group journeyed on, despite warnings by the infamous frontiersman Jim Bridger, who said the countryside was inhabited by rowdy Indians and grass was scarce and the timber heavy.

Brigham Young's wagon reached the summit of Big Mountain on July 23, 1847. From this summit he gazed for the first time upon the expansive valley below and uttered his now-famous words, "It is enough. This is the Place."

On July 24, 1847, the first company of the Camp of Israel, as it was called, finally entered the valley from the east, through Emigration Canyon. Today, residents of all faiths celebrate this historic event on the annual Pioneer Day holiday. City leaders erected "This Is the Place" monument at the mouth of Emigration Canyon to honor

Brigham Young, pictured here at the age of 65, became the leader of the Mormon Church following the murder of Joseph Smith in 1844. Young organized the successful emigration to Utah in 1846 and maintained his strong leadership in Salt Lake until the time of his death. Courtesy, Utah State Historical Society

those Saints who arrived that day and others who would follow.

At once the settlers began building their new empire. They diverted water from City Creek, planted crops, planned and drafted their city, and built homes, social halls, and places of worship (wards). Brigham Young immediately set aside several acres for the Mormon Temple. The unmistakable granite structure, with the gold angel Moroni blowing his trumpet from atop its spires, still graces the heart of Salt Lake City and is the apex of the valley's road system. Mormons are well known for their carefully planned communities. All streets extend north, south, east, and west in a matrix from Temple Square, the central block from which run South Temple, North Temple, East Temple, and West Temple streets.

Many early visitors were impressed with the layout of the city and commented on its clean, neat appearance. One visitor described it in 1850 as "a large garden laid out in regular squares." Mark Twain also wrote about the clean streams that trickled through town.

For a decade the Saints lived in peace in this Great Basin kingdom. Then, during Pioneer Day celebrations in 1857, Brigham Young received word of the approaching federal army. Young, who feared the army had been dispatched to drive them off and replace him with a non-Mormon governor, ordered the evacuation of the city. He instructed everyone—about 30,000 people—to fill their empty houses with straw and then move south. A few men remained behind to burn the city to its foundation if the federal army caused trouble. The troops, however, entered the valley and passed through the virtually deserted town without incident, heading on toward the Jordan River 40 miles south of the city where they established Camp Floyd.

Most of the people returned to their homes after the so-called Utah War, while others moved on to colonize other areas of the Utah Territory. History, however, does not ignore the significant impacts made by non-Mormons. Indeed, dynamic tensions between Mormons and non-Mormons fueled many of Salt Lake's early developments and achievements.

For example, beginning in 1865, gold and silver strikes in the canyons surrounding Salt Lake resulted in considerable economic development. Although Brigham Young was fiercely opposed to mining, the soldiers at Camp Floyd were not Mormon and were primarily responsible for most of the discoveries.

Mining and the approaching new transcontinental railroad—the Union Pacific and the Central Pacific—brought a tremendous influx of non-Mormons (gentiles) into the region. Economic confrontations erupted between Brigham Young and non-Mormon merchants. Mormons boycotted gentile-owned shops and, in 1868, they organized Zion's Cooperative Mercantile Institution (ZCMI), a producer's cooperative. Today ZCMI is a chain of full-service department stores and one of the valley's largest employers. ZCMI's main store still stands on Main Street in downtown Salt Lake City.

The first train of the Utah Central Railroad, pictured here in 1870, heralded the arrival of a new period of prosperity for the Salt Lake Valley. Courtesy, Utah State Historical Society

Salt Lake's rapid growth in the late 1800s is evident in this 1906 view of Main Street, which shows electric power, paved roads, and many businesses. Courtesy, Library of Congress

CENTER OF TRADE, CULTURE, AND COMMERCE

Salt Lake City soon became the capital of the Mormon Church. Church tithes were sent to Salt Lake City, where church leaders redistributed the funds for economic development.

In 1869 non-Mormons attempted to build a commercial capital of their own at Corinne near Brigham City and the new rail line. Brigham Young, however, blocked their attempts by enticing the two railroad companies to form the east-west junction at Ogden, where the line could be built south to Salt Lake City then west to California via the southern edge of the Great Salt Lake. Church officials convinced Mormon landowners to sell or donate their land to the railroad companies, which immediately accepted the generous gifts. Thus, Ogden thrived as the region's railroad hub, and Corinne's status quickly declined. With the new Utah Central Railroad, Brigham Young had effectively connected Salt Lake to the transcontinental railroad and, consequently, to the outside world.

The transcontinental railroad had two major effects upon the economic development of Salt Lake City: 1) merchants were able to market Salt Lake products, and 2) mining became a viable and extremely profitable industry. Naturally, Mormons welcomed these economic benefits, but at the same time feared the worldly influence of new immigrants on their conservative life-styles.

From 1869 to the end of the nineteenth century, Salt Lake City experienced considerable economic growth. Smelters were built up and down the valley to process the ores. Mining investors and businessmen such as Thomas Kearns, publisher of the *Salt Lake Tribune* and a U.S. senator, built elegant mansions along historic South Temple, east of the business district.

The city served as a hub for the non-Mormon mining towns and largely Mormon-

Salt Lake's local YMCA offered many social and cultural activities for the community. Here, a few young men take some quiet time for reading and a board game on the second floor of the YMCA Hall in 1905. Courtesy, Utah State Historical Society

Salt Lake's local YMCA offered many social and cultural activities for the community. Here, a few young men take some quiet time for reading and a board game on the second floor of the YMCA Hall in 1905. Courtesy, Utah State Historical Society

populated farming communities. Newspapers flourished, including the *Salt Lake Tribune,* established in 1871 as an anti-Mormon voice against the Mormon-owned *Deseret News.*

Although Salt Lake City had no formally organized political parties before the 1870s, political, social, and religious conflicts were common. Polygamy, practiced by many Mormons, became a major issue, pitting Mormon against non-Mormon. The controversy over polygamy culminated in the Edmunds Acts of 1882, which outlawed plural marriage. Realizing they could not outrun the federal government, Mormon leaders abolished polygamy in 1890 (a gesture partially responsible for Congress declaring Utah the nation's 45th state in 1896).

Salt Lake City, however, still faced other problems. Pollution from the many smelters and factories, health problems, and overcrowding due to the rapid growth and lack of a proper infrastructure diverted citizens' attention from the political battles of the late 1880s and 1890s. Citizens of all faiths and backgrounds were able to put aside religious differences and instead focus on issues beneficial to the entire community. One of the most important accomplishments of this unified effort was the creation of the Salt Lake Chamber of Commerce and Board of Trade. Its responsibilities were much the same as the Salt Lake Area Chamber of Commerce's are today—to promote business and industry and to foster a healthy economy in which business can thrive.

Episcopalians and Catholics responded to the city's health-care problems in the 1870s by opening St. Mark's Hospital and Holy Cross Hospital, which, for the most part, treated miners.

By the 1890s Salt Lake had become the financial and commercial capital of the West, but it was not immune to outside economic influences. The nationwide depression of the 1890s affected Salt Lake as much as any other city. Unemployment rocketed to over 40 percent, and the Mormon Church responded in many ways. The Church discouraged immigration into the city and encouraged citizens to join the Spanish-American War. The Church also financed several new businesses, including the onion-domed Saltair cultural resort along the shores of the Great Salt Lake. But construction came to a standstill during the depression of the 1890s. The only project to be completed during this period was the mammoth Romanesque-style City and County Building built on Washington Square in 1898. The building, which was recently renovated, is a showcase monument of the period architecture.

During the 1890s urban entrepreneurs modernized the city, installing electric lights and an electric railway system, but were slow to make other improvements, such as paving streets and laying water and sewage lines. The city also lacked proper health programs, and air pollution from smelters and railroad smoke increased.

In the 1890s Salt Lake City offered plenty of cultural activities. Citizens enjoyed musical and dramatic performances and attended sporting events and public debates. On Sunday afternoons families would gather for a picnic at a park or attend museum exhibits. Just as it is today, cultural enrichment was an important part of these early Salt Lakers' lives.

The turn of the century saw a decline in farming and an expansion in mining and commerce. Gold, silver, and copper poured from the hills. The Bingham Mine, under ownership of the Utah Copper Company (later known as Kennecott, and now called BP Minerals), employed more than 5,000 men in 1910.

A building boom in the first years of the twentieth century transformed the city into a sophisticated metropolitan area. Construction projects, such as the Mormon Church headquarters, the Cathedral of the Madeleine, and other religious edifices, accommodated the meteoric growth in population. As a government seat and educational center, the city built the state capitol building, office buildings, railroad depots, university structures, and the grand Hotel Utah. These stately structures still stand today, providing a pleasing balance of elegance and detail in older buildings with the sleek practicality of modern structures.

With careful planning and zoning, orderly growth, and progressive civic improvements, Salt Lake became a thriving city. Salt Lake City's 140,000 citizens encouraged public officials to participate in the "City Beautiful" campaigns common in American communities in the early 1900s. Residents planted trees and community leaders built parks, playgrounds, walkways, and boulevards. The beautification plan cost $1.5 million and probably rivaled any city's in the country. In 1928 Salt Lake City installed the

The Utah Copper Company, now known as BP Minerals, operated the Bingham Mine, pictured here circa 1915. The company employed more than 5,000 men at that time. Courtesy, Utah State Historical Society

Between 1935 and 1942, the Work Projects Administration (WPA) helped employ an average of 11,000 people each year in Utah. A number of these workers were hired to make mattresses and quilts, as shown here in this 1930s factory. Courtesy, Special Collections Department, University of Utah

The west side of the remarkable Salt Palace (foreground) is prominent in this view looking northeast toward downtown Salt Lake City. Photo by Steve Greenwood

world's first pneumatic-tire trolley buses, and in 1933 the city introduced rear-engine motor buses.

For the most part, prosperity continued until after World War I in 1919. Then a serious depression hit. When the Great Depression fell upon the rest of the world, Salt Lake was already in an economic slump with an unemployment rate of 35 percent, much higher than the nation's 25 percent unemployment rate.

The city, like the rest of the country, depended on Roosevelt's New Deal programs. These make-work projects continued through the 1940s, funding airport expansion projects and residential and highway construction. Utah then had the fourth-highest welfare rate in the country. Citizens with jobs came to the aid of their neighbors by taking pay deductions and participating in income sharing, donations, and volunteer fundraising programs. The Mormon Church, too, came to the aid of its poverty-stricken members with work programs and food.

Full economic recovery came when America entered the Second World War. Federal government agencies, particularly defense contractors, recognized Utah's ideal inland and central western location. To this day, Hill Air Force Base, Dugway Proving Grounds, and several private industries that manufacture radio tubes, electronics, and aerospace equipment continue to stimulate the area's economy.

As in most American cities during the war, crime increased. Fear of subversive activities was common. Warnings and prohibitions were given to Japanese-Americans in Salt Lake City as elsewhere, although the national headquarters of the Japanese-American Citizen's League was located here. Because of the war, more people participated in church and family activities and gained a better knowledge of current affairs and foreign countries.

During the war, Mormons and non-Mormons set aside most of their differences

"This Is the Place" monument commemorates the arrival of the Mormon pioneers to the Salt Lake Valley in 1847. Photo by Steve Greenwood

and were united in their concern for family, community, and country. With the Pioneer Day centennial approaching in 1947, the state legislature and the governor appointed a committee to oversee the erection of a monument to honor the city's founders. Mormon Church President Heber J. Grant chaired the committee with Vice-Chairman Duane G. Hunt, Bishop of the Salt Lake Diocese of the Roman Catholic Church.

Mining continued to fuel Salt Lake City's economy through the middle of this century. In 1950 *BusinessWeek* referred to Salt Lake as "the nation's greatest concentration of nonferrous mining, milling, smelting and refining." Kennecott was producing 30 percent of the country's copper during this period and was by far the area's largest private employer.

The area's burgeoning defense industry created extreme housing shortages. Residential construction finally met the housing demand in the 1950s, when people began abandoning the central city for the suburbs. As more people moved out of the central part of the city, community leaders became concerned with the encroaching urban blight. Architects and community leaders launched the Downtown Planning Association. The association's major accomplishment was drafting the Second Century Plan, a blueprint for revitalizing downtown by renovating and leasing vacant buildings and alleviating traffic problems.

After years of heated political debate, community leaders decided the valley needed a civic auditorium and arts complex. Participants in the long-standing debate finally agreed to build the Salt Palace at the center of the decaying city in hope of encouraging long-term urban redevelopment projects. Most economic and planning experts agree that this move was one of the most important planning decisions in the city's history.

A TALE BETWEEN TWO COASTS
In the 1970s Salt Lake City experienced a period of prosperity with the Overthrust Belt and its plans to develop oil-related resources. Industry flourished and the population swelled at an annual rate of 3.2 percent in the 1970s.

The glut of oil in the world market and the recession of the early 1980s put an end to the energy boom and left Salt Lake City, like most other regions between the nation's two coasts, in economic doldrums that prevailed well into the latter months of 1987. During the prosperous 1970s, immigration into Utah averaged 15,000 people annually. Then, in 1983 the state saw the first signs of emigration. Flat economic conditions prompted some 17,000 people to leave the state between 1983 and 1985. One report estimates that 2,700 people moved from Salt Lake County between 1985 and 1986. Officials at the State Office of Planning and Budget explain that migration trends are the most

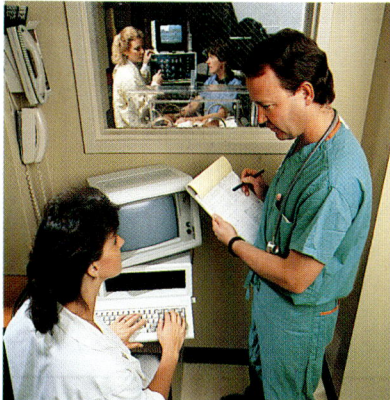

LDS Hospital is the development site for the Health Evaluation through Logical Processing (HELP) system, which uses computers and an extensive medical data base to improve the quality of health care in the Salt Lake Valley. Courtesy, Intermountain Health Care

volatile components of population forecasting; they believe this emigration is a short-term phenomenon. Most economic experts predict a gradual improvement in the area's economy, noting that the situation has "bottomed out," and there is a promising growth in the health services, finance, trade, and manufacturing areas. Most economists believe Utah has weathered the economic storm of the 1980s. "Salt Lake will soon see some brighter days," promises Jeff Thredgold, economist with Key Bank. One sign of an economic upswing is evident in the area's copper giant Kennecott. Kennecott, once the world's largest copper producer, closed its massive operations in Salt Lake County in 1985. Only a few of its 7,000 employees remained. Then, in 1987 Kennecott's parent company invested $400 million to modernize the plant. Today the firm, now called BP Minerals, is back in operation with more than 2,000 employees.

As with most of the nation, Salt Lake's economic base has shifted from agriculture and mining to the service sector: trade, government, finance, insurance, real estate, transportation, communications, and public utilities. Utah is one of only two states in the Rocky Mountain region reporting employment growth. Furthermore, of all the states in the Intermountain West, only Utah and New Mexico show any signs of job expansion. When compared with the rest of the country, Utah's 2.21 percent increase in employment looks favorable. It is estimated that 570,000 new jobs will be added in the next 20 years for a total of 1,188,000 in 2010. Utah is still a growth state, with population and industry growth rates predicted to average almost two percent through the remainder of this century and well into the next, about twice the national average.

MODERN PIONEERS

The same determination and optimism that drove a band of settlers to this land in 1847 still exists in Utahns today. The University of Utah's Health Sciences Center is the engine fueling much of the pioneering research and technology being developed in the valley. In 1982 Dr. William DeVries implanted the world's first artificial heart into Dr. Barney Clark. A dentist from Seattle, Clark lived 112 days with the apparatus. A true pioneer, Clark gave his life in an experiment that will allow others to live.

Patrons of the Market Street Broiler enjoy a delicious Saturday lunch. Photo by Stephen R. Smith

Salt Lake Valley's most valuable resource is its increasing population of young people. These children relax in Liberty Park in Salt Lake City, which occupies six square city blocks and provides a quiet refuge from the bustle of the city. Photo by Stephen R. Smith

The artificial heart, invented and developed in Salt Lake City, is the latest in a lifetime of accomplishments for Dr. Willem Kolff, a native of the Netherlands who came to the University of Utah in 1967. When searching for a place to further his artificial organ research, Dr. Kolff looked "for a place where there could be the possibility of a cooperative program between the various colleges, specifically medicine, engineering and pharmacy," he explains. Known as the "Father of Artificial Organs," Dr. Kolff invented and perfected the artificial kidney that today benefits more than 250,000 people worldwide. He also pioneered work on heart-lung machines and advanced the kidney transplantation process. His achievements have attracted unprecedented attention to the University of Utah and the new "Bionic Valley," as the Salt Lake Valley has become known. Dr. Kolff retired in 1986, at age 75.

Indeed many Salt Lakers are responsible for today's news-making technology. For example, engineers at Colmek Systems Engineering designed and built the underwater sonar technology used in locating the *Titanic* at the bottom of the Atlantic Ocean in 1985. Scientists also used the Jason Jr., a remote-control robot, to explore and film the interior of the giant vessel. Salt Lake technology was also on board the famous *Voyager* flight. Engineers at Hercules helped design the *Voyager* aircraft, a twin-engine lightweight graphite plane capable of carrying five times its own weight in fuel. The pilots and aircraft set new aviation records in December 1986 when they flew 27,500 miles around the world in 11 days without refueling. Hercules donated graphite, high-tech equipment, facilities, technicians, and labor to the enormous project. Much of the aircraft was built during two years at the Hercules plant in Magna, Utah, located in Salt Lake County.

These are but a few of the many examples of the talents, skills, and resources that entrepreneurs are developing in Salt Lake City today. Other innovations include lasers, holograms, and computer graphics.

Once considered a remote region, safe from the rest of the country and the scrutiny of the federal army (the nearest town was more than 500 miles away), Salt Lake City has gradually earned a reputation as the "Crossroads of the West." Living up to its nickname as The Gathering Place, the city today is a center of trade, culture, and commerce.

"During the past 40 years, entrepreneurs from around the nation and world have come to Utah to build components for our national defense and space efforts, to develop artificial organs and even to establish headquarters for a chocolate chip cookie em-

The warm rays of the sun highlight Utah's capitol building, which sits immediately north of downtown Salt Lake City at the top of a steep hill that overlooks the valley. Photo by Stephen R. Smith

pire," says Governor Norman Bangerter. "We welcome these entrepreneurs whether they were born here or came here to realize their dreams."

SALT LAKE CITY TOMORROW

As Salt Lake City races with the rest of the world toward a new century, it will face new challenges and new promises. John Naisbitt, author of the best-selling *Megatrends,* challenges Salt Lake City's destiny, calling it "one of ten new cities of great opportunity." *Town and Country* magazine labels it "Vibrant, Vigorous and Still Visionary." *The Christian Science Monitor* describes it as one of the nation's nine cities on the rise, a "land of high achievers and plain dealers, full of self reliance." This optimism is not simply news copy. The city has continued to thrive since its first pioneers stepped foot on its baked soil. The Salt Lake-Ogden Metropolitan Statistical Area (MSA) is predicted to grow 30 percent between 1988 and the year 2000, while the nation will grow only 14 percent. During the period 1986 to 2010, a total of 187,000 net immigration is expected to occur in the state.

Most people visiting or moving to Salt Lake are immediately taken with the idyllic scenery along the Wasatch Front; indeed, the life-style here is enviable. Yet as beautiful as this all is, there are drawbacks—as with any area. The Wasatch Front, the name given to the urban corridor that stretches from Salt Lake to Ogden, has a large population of young people. And in the next decade, residents will have to face the problem of how to maintain high standards of education in an overburdened school system. "Utah's public education system faces several more years of increasing demands and needs, but Utah's taxpayers may be at or near their capacity to give," concludes a report prepared by the Education Reform Subcommittee of the Salt Lake Area Chamber of Commerce.

However, when the state's school-age population, which is 25.5 percent of the state's total, eventually moves into the labor force—increasing the number of available workers by 32.2 percent by the year 2000—this supply of young, educated people will be an

The Jordanelle Dam, under construction one half-hour east of Salt Lake City, will serve the area's water needs in the coming century. Photo by Stephen R. Smith

asset to the state and will be far more valuable than the cost of educating them today. The 1987 legislature realized this and approved a $160-million tax increase, the largest in the state's history. Most of the revenue went to fund public education, which citizens recognize as a crucial investment in the community's future. Companies are already looking to Utah because of its young, bright, regional work force.

Unknown to most people who have never visited the area, Salt Lake Valley is one of the most urbanized areas in the entire country. Salt Lake also has its share of problems that go along with urbanization. Two of the city's problems are due to geography.

Salt Lake City is situated in a mountain valley. In winter, especially January, Utah's usually clear blue skies and sunny days ocassionally turn to a cloudy haze. The inversion, as the climatologists call it, is a temporary condition in which a haze settles over the valley like a lid over a frying pan, trapping cold, moist air from the Great Salt Lake beneath it until a good breeze or storm washes it away. To escape these gray winter days, residents can drive up the nearest canyon road for a daily dose of warm sun,

clear skies, and deep snow—a never-fail cure. But inversion is a problem, and scientists at the University of Utah are researching ways to solve the smog problem by cloud seeding the air with ice crystals and liquid carbon dioxide, which clears the fog and allows cleaner air to move in. At present, cloud seeding is the best and only proven method for killing fog. Although this method has been used at airports in winter in many parts of the world, Salt Lake City would become the first city to carry out such a project.

As the second driest state in the country, Utah also must continue to find resources to supply enough water to its growing population. The Jordanelle Dam, currently under construction on the Provo River 40 miles southeast of Salt Lake City, will substantially increase the water supply for the

Total construction in Utah from 1984 through 1986 totaled more than one billion dollars per year. Salt Lake's skyline changed dramatically during this time and continues to experience growth. Photo by Stephen R. Smith

Wasatch Front. The reservoir's 370,300 acre-feet of water storage will be complete in about six years and is being called the West's last major dam project.

To effectively handle future transportation needs, planners and engineers have suggested a three-point plan. The plan includes a combination of widening existing freeways, adding on-and-off ramps, and installing a light-rail transit system along the existing Union Pacific right-of-way tracks which parallel Interstate 15.

Community leaders and residents are working hard to avoid the problems other cities have experienced because of explosive growth. Planners are aware that a healthy growth in business also means population problems and inevitable urbanization. With a new generation of bold young minds at the helm, Salt Lake stands ready and willing to stride into the future, yet is carefully planning each step.

Salt Lake City Mayor Palmer DePaulis' blueprint for building the city's future addresses six issues for the coming years: 1) neighborhood vitality, 2) downtown vitality, 3) economic development, 4) government capacity and structure, 5) human development, and 6) leadership. The success of Salt Lake City Tomorrow Project, as the plan is called, depends on creating new communication channels among politicians, educators, corporate heads, retailers, religious leaders, urban designers, and other community leaders.

Economic development programs have taken on a sophisticated tone in recent months. The newly formed Utah Economic Development Corporation (UEDC) has assumed sole responsibility for conducting unified economic development activities for Salt Lake County. This nonprofit organization's primary mission will be to attract new businesses and jobs to the area. This coalition of representatives from the various municipalities in Salt Lake County and private individuals has resulted in several business relocations and has eliminated turf battles for new industry within the county and, many times, within the state.

Another community action program is Project 2000, a nonprofit organization dedicated to examining issues and changes facing the state. Salt Lake's NBC television affiliate, KUTV, sponsored a series of documentaries which explored issues such as population changes, education, and transportation. Recently a group of 17 community leaders (including businesspeople, professors, publishers, a Utah Supreme Court Justice, state legislators, and others) created the Coalition for Utah's Future. This group will help set Utah's future agenda and will encourage and support elected leaders who support the coalition's agenda.

The Mormon Church, headquartered in Salt Lake, is an influential organization with a large following in the state. The state's population is approximately 70 percent Mormon and 30 percent non-Mormon, although in the metropolitan area of Salt Lake the population is closer to 50 percent non-Mormon. Church leaders are concerned about the well-being and success of the community, as are other citizens, and agree that the community needs to encourage leadership from all groups. Richard Lindsay, managing director of public affairs for the Mormon Church, says:

Utah needs a cross section of committed leaders from all areas of the business, civic, religious, ethnic, academic, governmental and labor sectors. Members of the dominant church and all other churches need not view this as a special problem, but seek to build bridges of brotherhood in all formal and informal community structures, and make this religious diversity the basis of enriching growth for both groups.

Salt Lakers must ensure their children a prosperous future and guarantee them the same standards of living Utahns enjoy today. The members of the Coalition for Utah's Future "envision a Utah future rich in opportunity and quality of life. To realize this vision, we pledge our best efforts to achieve: economic development and job opportunities, quality education, superior cultural and physical environment, recognition of the strengths of diversity consistent with traditional values. This is our promise to Utah's future generations."

"We're not leaping into the future pell mell," says Mayor DePaulis. "We're managing our growth. We won't kill the goose that keeps giving us golden eggs."

Above: The headquarters of the LDS Church, seen here during the early summer months, is located at the heart of Salt Lake City. In this view, looking down on the main court area, the Salt Lake Temple is visible at the back of the court with the Church Office Building visible to the right. Photo by Steve Greenwood

Left: A picturesque sunset over the Great Salt Lake is enjoyed by this young family. Courtesy, Utah Travel Council

CHAPTER TWO

WELCOME TO THE NEIGHBORHOOD

The Wasatch Mountains that Twain admired have not changed much in the last 120 years, and, in some respects, neither have the reasons for living here. The people who live here today still exemplify the industriousness which gave the state its nickname: the Beehive State. They hold dear the virtues of their pioneer heritage: honesty, trust, pride in the community, commitment, hard work, cooperation, family, friends, health, conservative ideals, and clean, peaceful living. These virtues have attracted people and industry to the Salt Lake Valley.

Moreover, it is impossible to visit or live in the Salt Lake region without noticing the magnificent beauty that surrounds it.

THE NATURAL SETTING

Salt Lake County is nestled in a mountain valley at 4,200 feet above sea level. Two mountain ranges, the Wasatch Mountains to the east and the Oquirrh Mountains to the west, surround the northern and southern part of the valley, forming the county's boundaries. Below the mountains lies the Salt Lake Valley, an oasis of pastoral greens surrounded by steep, cool canyons and cascading streams. The gleaming skyline of the city is dwarfed by this mountainous backdrop, which changes color from deep purples and shimmering greens in summer to blazing reds in autumn and rosy white in winter. To the northwest lies the expansive Great Salt Lake, the world's largest inland sea. Beyond that extends miles and miles of sage, sand, and rock.

THE PEOPLE

Nearly 60 percent of the state's population resides in the Salt Lake-Ogden metropolitan area of Salt Lake, Davis, and Weber counties. About three-fourths of all Utahns live along the metropolitan corridor known locally as the Wasatch Front. The Salt Lake metropolitan area is the 37th largest metropolis out of 281 metropolitan areas in the country, according to the U.S. Bureau of the Census. In fact, between 1980 and 1986 Salt Lake-Ogden was the ninth fastest growing Metropolitan Statistical Area with a population of more than one million. Since 1980 the population has increased 14.4 percent.

Today Salt Lake County (which includes Salt Lake City) is home to more than 700,000 people, who have chosen to live here for a variety of reasons. Some came to ski "the greatest snow on earth" and decided to find a job and stay. Others came to attend one of the area's outstanding universities or colleges. Still others came because it is a great place to raise a family. People are attracted to the area because of the good schools, the honest work ethic, the breathtaking scenery, and the variety of recreational opportunities.

In fact, Utah is considered one of the five most livable regions in the country. A recent Cleveland State University study rates Utah extremely desirable for retirees, racial minorities, and business leaders. Although quality of life is measured differently by everyone, a city must meet the basic needs of its citizenry. According to Joel A. Lieske, associate professor of political science and conductor of the Cleveland State study, "In general it is better to be rich than poor. It is better to be safe than insecure. It is better to be well served by government than poorly served. It is better to live in a clean environment than a dirty one. And it's

The majestic Wasatch Mountains tower over the glowing skyline of Salt Lake City. Photo by Stephen R. Smith

This stunning view of the Great Salt Lake and Antelope Island was taken from the Wasatch Mountains to the east of Centerville. Photo by Steve Greenwood

A mother and her children wade in the briny surf of the Great Salt Lake near the Saltair Resort. The Saltair is now undergoing reconstruction after suffering flood damage. Photo by Steve Greenwood

Above: Salt Lake features some of the best skiing in the country. Photo by James W. Kay

Left: The majestic Utah State Capitol is enhanced by the sight of blossoming cherry trees each spring. Photo by Steve Greenwood

A new mother takes a quiet moment with her baby in a homelike postpartum room at the Center for Women's Health at Cottonwood Hospital Medical Center in Murray. Courtesy, Intermountain Health Care

better to live in a community rich in amenities than one that is poor." Indeed, Salt Lake is rich.

As major metropolitan centers go, Salt Lake is a fairly young city—only 140 years old—and has a unique blend of western hospitality, eastern sophistication, and midwestern work ethics. Its population is also young. The median age is 25.5 years, the youngest in the nation. One half of Utah's residents are younger than 25.5, which is 6.2 years younger than the national median age of 31.7. Between 1975 and 1985 Utah's public school-age population increased by 97,000, an increase of 30 percent. During this same period the nation's school-age population decreased by 6 million, a decline of about 12 percent.

This tremendous growth in population is unique to Utah. The Mormon Church, with its international headquarters in Salt Lake City, encourages large families, and Utah's birthrate reflects this belief. At twice the national average, the state's birthrate

The town of Ogden, with the Great Salt Lake visible in the distance, is located about 45 miles north of Salt Lake City. Photo by Steve Greenwood

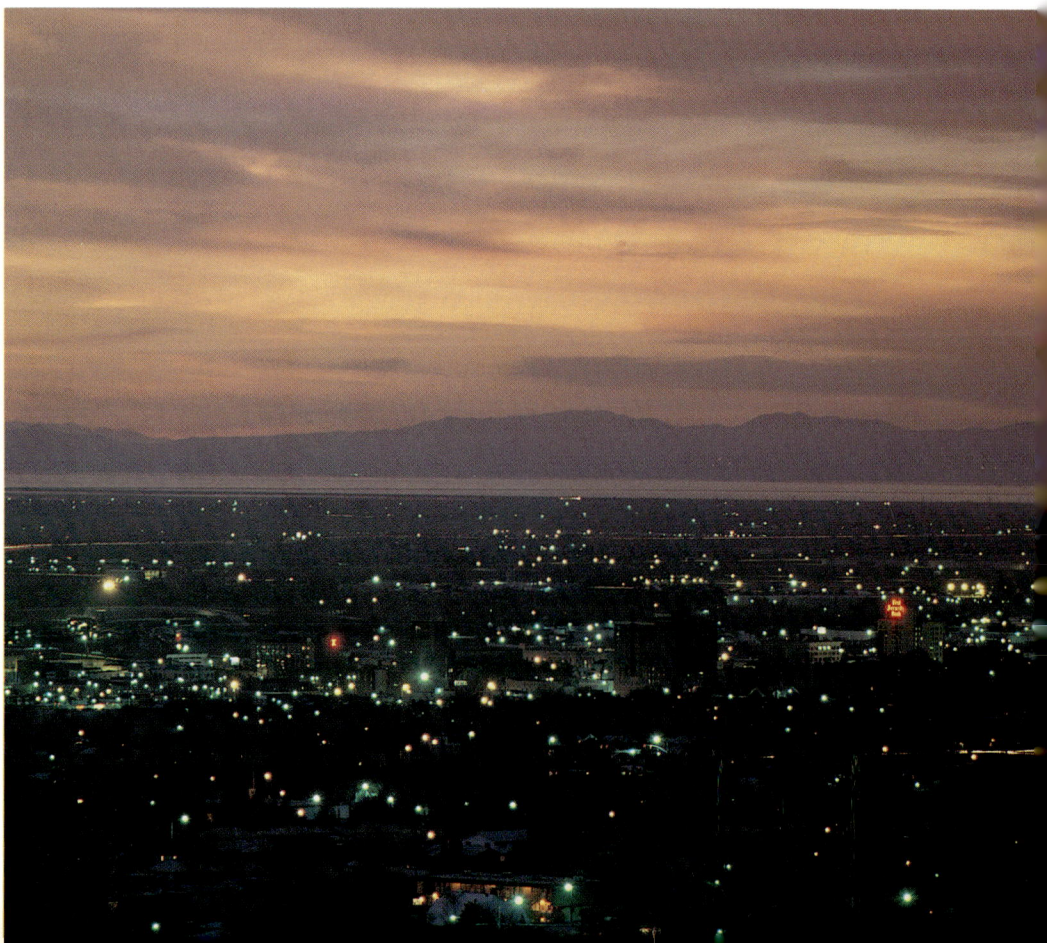

jumped dramatically between 1976 and 1982, averaging 3.2 births per woman. In recent years the birthrate has decreased to 2.6 births per woman and is expected to decrease further—2.5 births by 1990. The lower birthrates will not decrease the school-age population until well into the 1990s, and business and community leaders are working together to maintain high educational standards without raising taxes.

Salt Lakers have proven their ability to adapt to change. After a few years of slow economic growth, Salt Lake City is once again buzzing with energy. Residents are again enthusiastic and optimistic about their city's future. They have lived through the rough times and have transformed hardships into positive, creative action.

HOUSING

Within Salt Lake City, 12 municipalities function as distinct cities with mayor-council forms of local government. Politically, Utahns tend to be nonpartisan conservatives who traditionally switch parties every 15 to 20 years. Regardless of political affiliation, however, Utahns love their state, and their passion is contagious. Their revived energy has replaced doubt, and today they are again striking that spark of independence and showing the self-reliance and optimism inherent in Utah's heritage. Each of the 12 municipalities offers affordable housing, quality schools, and convenient shopping. Because of low inflation and stable interest rates, the price of housing in Salt Lake City has stabilized during the 1980s. In 1988 the average price of a house was $76,602, a decrease of 1.18 percent from 1987. Recently, new housing construction has been highest in the southern and western quadrants of the valley: Sandy, West Jordan, West Valley City, and South Jordan.

Houses are available at reasonable prices throughout the valley. According to the American Chamber of Commerce Researchers Association's (ACCRA) Inter-City Cost of Living Index, the average price of a new house with 1,800 square feet of living space in a mid-management level neighborhood is around $87,000. With a housing index of 94, Salt Lake's housing costs are well below the national average of 100. The panoramic views of the valley seen from homes in established neighborhoods on the north and east benches (the Avenues, foothills, and Mount Olympus Cove areas) command somewhat higher prices.

The Salt Lake area offers a variety of homes to meet everyone's tastes. Winding

Salt Lake Valley offers ample room for its continuing development and for its increasing population. The Wasatch Mountains create a natural backdrop in this expansive view, looking southeast from downtown Salt Lake City. Photo by Steve Greenwood

33

up the steep hills immediately north of downtown, the Avenues district offers large, historic, and executive-style homes with a view of the entire valley. In town, condominium projects such as Governors Plaza, American Towers, and Eagle Gate offer luxury living. Also close to town is Rose Park, a quiet neighborhood with quaint houses and tree-lined streets. Further out, located at the southern end of the county, is Sandy City, one of the fastest-growing municipalities in the country; it is also the gateway to canyon roads leading to first-class ski resorts such as Alta, Snowbird, Brighton, and Solitude.

Draper City, a more rural neighborhood, is located at the extreme southern end of the county and offers acres of land for a ranch, orchards, gardens, or prize horses. Although this is country living, it is a mere 40-minute commute to downtown Salt Lake City.

Apartments are plentiful and also quite reasonable. According to an ACCRA survey, an average monthly rent for a two-bedroom apartment is $332. And the average price for a condominium is $68,600.

HOW SALT LAKE COMPARES

According to an ACCRA survey, which compares costs of 27 grocery items, utilities, transportation, health care, and miscellaneous items such as entertainment, Salt Lake's Cost-of-Living Index of 99.3 is slightly below the national average. Groceries in Salt Lake, at 95.4 on the index, are roughly 5 percent less expensive than elsewhere in the nation, about 8 percent less than in San Diego and 15 percent less than in Dallas.

Utility costs, at 96.2 on the index, are roughly 4 percent below the national average. The average monthly fuel bill runs about $50 per month. The average monthly electric bill is about $53 per month.

Property taxes in the Salt Lake area are quite low, averaging between 1.5 and 2 percent. Property taxes are assessed at market value less 20 percent statutory reduction and less 25 percent for a residential exemption (for primary residences). For example, the annual property tax on a $100,000 house would be $930.

Utah's individual income tax is 7.75 percent for all income over $7,500. The state

Quiet residential neighborhoods are found throughout the Salt Lake Valley. Lamp posts, lush trees, and neatly groomed yards line the streets and avenues. Photo by Stephen R. Smith

Snow-capped Mount Olympus provides a spectacular view for the residents of the Foothill Place Apartments in Salt Lake City. Photo by Steve Greenwood

The stately Mormon Temple, located at historic Temple Square in Salt Lake City, was built in the early days of the Mormon Church. Temple Square is one of the most visited tourist attractions in the state of Utah. Photo by Steve Greenwood

sales tax is 6.25 percent. Average annual taxes for a family of four with a household income of $25,000 in 1987 were $2,419.00.

MEET ME AT THE MALL

Six regional malls throughout the county make shopping a breeze on weekends or after a busy day at work in town. Cottonwood Mall in Holladay (in the southeastern section of the valley) was one of the nation's first indoor malls. Recently remodeled, the mall is a mosaic of modern interior design and has two major department stores, ZCMI and J.C. Penney, as well as numerous smaller shops.

Valley Fair Mall is in the western half of the city and was recently renovated; it has over 100 stores including ZCMI, Mervyn's, and J.C. Penney. Fashion Place Mall serves the rapidly growing southwestern section of the valley with more than 100 shops and major stores, including Sears, Nordstrom, and Weinstocks. The valley's newest mall, South Towne Center, serves the extreme southern end of the valley, Sandy and Draper, with dozens of retailers and ZCMI's largest store.

Tourists, convention-goers, and downtown office workers need only to cross the street from either their hotel rooms or their offices to shop or browse through two huge indoor malls. Crossroads Plaza and the ZCMI Center, both located on Main Street, offer hundreds of shops in open atriums, plenty of parking, theaters, and dozens of restaurants.

Several other neighborhood shopping centers, such as Brickyard Plaza, Foothill Village, Creekside, and Hillside Plaza, have spacious supermarkets, drugstores, hardware and home improvement centers, theaters, and sporting-goods stores.

PLACES OF WORSHIP

No community in the country better recognizes the importance and right of practicing one's religion in peace than Salt Lake. After all, freedom of religion was the principle upon which the valley was first settled 140 years ago. Today all major religions are practiced in the valley.

Of all the major religious structures that were built in the early days of the city's settlement, perhaps the most famous is the Mormon Temple. The Cathedral of the Madeleine, on historic South Temple east of downtown, attracts Catholics and visitors from around the country. The cathedral is an architectural wonder with spectacular stained-glass windows, massive columns, and intricately detailed, handcarved sculptures. Other historic places of worship include the Holy Trinity Greek Orthodox Church, First Presbyterian Church, First United Methodist Church, and the Kol Ami Synagogue.

Several modern structures also exist throughout the county.

Rabbi Frederick L. Wenger of the Congregation Kol Ami moved his family to Salt Lake because "it has a well-deserved reputation as a place where faith is taken seriously, where education is valued, where families are appreciated and where all children are seen as an unmixed blessing."

A SAFE PLACE TO LIVE

As one of the most livable cities in the country, Salt Lake City is also a safe place to live. The 1985 Salt Lake metropolitan area crime index was 6,425 per 100,000 population, much lower than other urban centers. Of this total, 337 were violent crimes. The remainder were all crimes against property. Utahns are also more likely to report crimes than residents in other states. Bare figures are misleading, because they include crimes as serious as homicide and as minor as hubcap thefts. According to the Federal Bureau of Investigation, the actual crime rate in Salt Lake is about one-third that of other major metropolitan areas on the East and West coasts.

Commended in 1986 as one of the best in the country, the Salt Lake Police Department is progressive and efficient in solving cases, according to an outside consultant. "The community should be proud of the department and its leadership."

As Ted Wilson, Salt Lake mayor from 1974 to 1984, writes in Utah's *Wasatch Front*, "American cities are plagued by sameness. People and natural surroundings distinguish one city from the next. Salt Lake, surrounded by scenic splendor and filled with friendly people, is a city above the rest."

That's also why Robert Redford, famous actor, director, and environmental spokesperson, lives in Utah. "When you have a gem like this particular state and the common decency of the people of this state, it's in our interest to preserve it the best way we can," Redford says. "If we are to raise children in this day and age, we want to guarantee something of a future for them. The state has an innocence about it, a vitality about it. It has space. It has good space . . . It has space to grow in."

Mark Twain would agree.

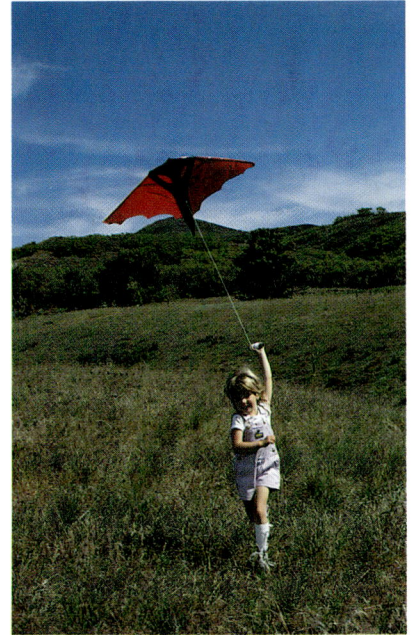

This young girl delights in an afternoon of kite-flying in the foothills near Salt Lake City. Photo by Steve Greenwood

Salt Lake has been praised as a vital and friendly area, distinguished by beautiful scenery and abundant recreational opportunities. Photo by James W. Kay

SALT LAKE MEANS BUSINESS

When it comes to business and research, Salt Lake City is as modern as any city in the country. Salt Lake leads the nation in revenues allocated for education. It is third in the nation (behind California and Massachusetts) in the number of start-up companies based on university research, and it is recognized throughout the world for its leadership in biomedical science, space engineering, communications, and other technologies. Salt Lake City is hardly behind the times.

Salt Lake businesspeople may have different reasons for their enterprising ways, but on one thing they agree: They love doing business here.

"In Salt Lake, people have a tremendous, family-driven need to succeed that translates into a profitable business," says Merlene Leaming, executive vice president and part owner of Clark Leaming Designs for Business. "Other major cities are saturated. In Salt Lake, it's easy to be recognized for your talents and for a job well done." Leaming credits the success of her 35-year-old interior design business to Salt Lakers' discriminating tastes. "The city gave us an educated clientele, people who know the difference in good design and quality."

Many people believe the Mormon Church and the University of Utah are responsible for much of the consumer sophistication found in Salt Lake. When June Morris opened her one-person travel agency in 1970, she had no idea her business would one day grow to become one of the nation's top 15 travel companies and the fastest-growing woman-owned business, doubling its sales volume year after year. Morris Travel and its charter affiliate Morris Air Service have achieved meteoric success largely because of its Salt Lake City location. "The Mormon Church has a positive influence on our industry because missionaries travel throughout the world," Morris says. "Families are aware of other countries and other cultures. I think it's a dynamic city, eager to know and learn new things."

Indeed, it is this hunger for new technologies and services, combined with an appetite for basic manufacturing, that makes Salt Lake's economy so diverse. Although the Salt Lake metropolitan area is one of the nation's top 10 fastest growing urban regions by population, the area's business climate fosters growth of its own. *Inc.* magazine consistently ranks Utah among the top 15 states in its annual "Report on the States." The rankings, according to the magazine, "measure how a state is actually doing in stimulating entrepreneurial activity and economic expansion. A state's position on the list reflects its economy's relative success, over a four-year period, in three areas: job generation, new business creation, and young company growth."

In his best-selling book *Megatrends*, John Naisbitt describes Salt Lake City as a "megatrend" city, one of 10 areas in the country with promising economic growth and opportunity. Salt Lake is in good company with Phoenix, Denver, St. Louis, Boston, Tampa, San Diego, Memphis, Kansas City, and Portland. "Businesses and entrepreneurs looking to expand or relocate would do well to look closely at these cities—the likely stars of the 1990s," Naisbitt writes.

Other accolades have been showered upon Utah's capital city, and they all point to one distinction: It is one of a dozen or so cities expected to lead the nation in economic growth. A nationwide survey of 2,000 banking chief executive officers declares Salt Lake City one of the 10 western "Cinderella Cities."

Computers, holograms, lasers, mutual

Salt Lake City's central business district is compact, occupying approximately eight square blocks. Photo by Stephen R. Smith

Beneficial Life

First Interstate Bank

Salt Lake Sheraton

VALLEY BANK & TRUST

Above: The glass-adorned office towers of Salt Lake's dynamic business community blend with the older, more traditional structures of the area. Photo by Mark Gibson

Above right: The Eagle Gate Plaza and Office Tower is the premier office structure in the Salt Lake Valley. Completed in 1986, it is the newest addition to the growing Salt Lake City skyline. Photo by Steve Greenwood

funds, traveler's checks, semiconductors, rocket motors, and artificial limbs are just a sampling of the kinds of goods and services offered by Salt Lake Valley companies.

Salt Lake City has attracted some of the most prestigious and vital companies in America, including Hercules Aerospace, Unisys, National Semiconductor, Litton Guidance and Control Systems, Delta Air Lines, American Express, Fidelity Investments, and McDonnell Douglas, to name a few. Some of these firms relocated to Salt Lake; others, such as Evans & Sutherland, designers and manufacturers of state-of-the-art computer graphic systems and flight simulators, are homegrown companies that flourish along the Wasatch Front. These companies have discovered that they can be more competitive in Salt Lake than in many other cities around the country.

Part of the reason is cooperation. Local government is extremely responsive to the business community in Salt Lake. According to McDonnell Douglas Aircraft Company officials, "The reception we've received in Utah has been fantastic in every aspect. Every place we turned, we've gotten help far beyond our expectations. The cooperation of government and business leaders has been excellent."

Utah's legislature has historically been responsive to business concerns. Despite fiscal shortages and an already small tax base, certain industries benefit from tax incentives, such as sales tax exemptions on manufacturing equipment and machinery. However, some critics are concerned that business is not paying its fair share of taxes in Salt Lake.

Salt Lake City is at an economic crossroads, and community leaders and lawmakers agree they must create a climate to foster business growth from within and outside the valley. "There are a lot of factors out there that are changing, and we need to change with them," says Joseph A. Cannon, president and chairman of Geneva Steel. "Utahns must recognize opportunity in the state and act on it." Although not in Salt Lake County, Geneva is a good example of the keen resourcefulness Utahns have for

Marketing, engineering, and research and development firms are among the wide variety of businesses which help create a strong economy in the Salt Lake Valley. Courtesy, the University of Utah Research Park

business. When USX Corporation announced its plans to close their 50-year-old steel mill located in Orem, a few miles south of Salt Lake City, Cannon and his partners decided to buy the plant because they believed the area was on the verge of tremendous economic growth. Cannon, a native Utahn working in Washington, D.C., saved about 1,500 jobs with the purchase.

Economic development is more than just a catchy phrase in a press release. Community leaders realize that a strong economy requires cooperation between the private and public sectors. In an effort to attract new business, encourage local expansion of existing firms, and create jobs, the governor, mayors, chambers of commerce, and the private sector have worked together to apply more than lip service to the city's development. The result of this unity is the Utah Economic Development Corporation (UEDC), a partnership of private and public interests responsible for spearheading economic development programs in Salt Lake County. UEDC's board of trustees includes the valley's 12 mayors, 3 county commissioners, the executive director of Utah's Department of Community and Economic Development, and 16 representatives from the private sector. "This is an unprecedented coalition of public and private individuals and groups who recognize that greater economic development holds the key to our continued growth and development," says D.N. "Nick" Rose, cofounder and chairman of UEDC. Rose is also president and chief executive officer for Mountain Fuel Supply.

This promising coalition will be vital to the valley's future success. "The cooperation of the state of Utah, Salt Lake County, each of the cities, the chambers of commerce, and private enterprise will play a major role in stabilizing the economy of our valley and state," says County Commissioner Bart Barker. "We must all take an active role in telling the Utah story and let people know we are a viable, growing, and exciting place in which to live, work, and play."

The Salt Lake Area Chamber of Commerce is committed to helping existing businesses expand and make a profit. Since 85 percent of the state's new job growth comes from small business firms, the Chamber has designed programs through its Government Affairs and Small Business councils that will promote new and existing small business enterprises. The Government Affairs Council focuses on increasing profits and creating jobs for small businesses. "Every action taken by government directly affects the economic

Salt Lake's rapidly growing skyline features many modern high rises. Photo by Stephen R. Smith

Above: Salt Lake's central business district has expanded dramatically in recent years. Photo by Stephen R. Smith

Right: Salt Lake City, the hub of the Salt Lake Valley, is also the financial, commercial, and transportation center of the Intermountain West. Photo by Stephen R. Smith

strength and future of business in Utah," says Gary Fisher, chairman of the Chamber's State Legislative Action Committee.

The Chamber's Leadership Utah program is one of the most successful programs the organization has ever undertaken. This program offers training to the valley's future business leaders and covers community issues related to education, religion, state and local government, economic development, the arts, transportation, and crime and law enforcement.

Another invaluable resource for Salt Lake businesses is the Utah Small Business Development Center (USBDC). Headquartered at Research Park in Salt Lake, the center has offices downtown and branches throughout the state. Founded in 1979 to help reduce small business failure, the USBDC assists entrepreneurial activity, encourages sound business start-up activity, promotes timely and proper business expansion, and assists businesses that are unique and beneficial to the region.

Perhaps most important of all is the exceptional quality of the work force. "Salt Lake has one of the best, most professional labor forces in the country," says Dino Georgalas, general manager of the Salt Lake Marriott Hotel. Born in Cairo, Egypt, Georgalas was reared in South Africa. He has worked in cities in Canada, Switzerland, and Denmark, as well as in Chicago, Detroit, and Los Angeles. When the Marriott Corporation built its Salt Lake hotel in 1981, he requested a transfer to Utah. A worldly, well-traveled man, Georgalas finds the people of the Western Rockies to be warm, genuine, and friendly.

Inc. magazine's 1985 "Report on the States" ranked Utah's labor market second in the nation. "The labor ranking weighs what a state 'gets' in terms of productivity (dollar amount added by each employee to the raw value of manufactured goods) and a more sophisticated work force (percent of workers older than 25 who are high school graduates) against what it 'pays' in terms of higher wages and extent of unionization."

Salt Lake's most valuable asset is its people. While much of the nation is facing a possible labor supply shortage, Salt Lake's work force is increasing. Some project that by the end of this century, the city's work force will increase by 30 percent. Women will account for more than half of this increase. It is estimated that by the year 2000, 62 percent of Salt Lake women will be part of the workplace, a 59 percent increase in female workers. With Salt Lake's growing role in the international market, the city realizes it must prepare itself, its industries, and its residents to compete on a global scale.

Salt Lake's economy is in transition, and the winds of an economic renaissance are blowing. Heavy industry is no longer the heart and soul of Salt Lake's economy, but

Salt Lake Valley's energetic work force has achieved an impressive reputation for its professionalism, loyalty, and unflagging work ethic. Photo by Steve Greenwood

it still plays an important role. The reopening of two of the area's once-largest industries, BP Minerals (formerly Kennecott Copper) and Geneva Steel, signals this renaissance. The reopening of these plants—and the opening of many new ones—is a promising sign that will likely carry the community into a new era and provide a solid foundation upon which to build new industries and employ more people.

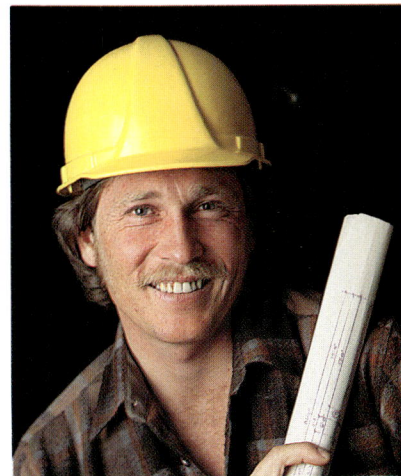

Salt Lake's most precious resource is its people. Photo by Steve Greenwood

MANUFACTURING: MAKING HISTORY

In 1987, when much of the nation between the two coasts experienced plant shutdowns, layoffs, and farm closures, Salt Lake City held its own. Not isolated from these national trends, it had its share of downturns, of course. It seemed when one company announced it was expanding, another would announce cutbacks. Yet, Utah was one of only two states in the Rocky Mountain region reporting employment growth.

According to an economist in Denver, Salt Lake City's economic growth will occur mainly in the areas of health services, finance, trade, and manufacturing industries. She predicts this growth will average about 2 percent per year in the coming years, with each percentage point equaling 6,400 jobs.

Manufacturing has historically comprised about 15 percent of the area's industrial base. From 1973 to 1986 the number of jobs in manufacturing in the West rose 10 percent, while the number of manufacturing jobs in the rest of the country declined by 9 percent. Utah is especially strong in defense-related aerospace manufacturing, according to a Federal Reserve Bank newsletter. In the West, manufacturing employment growth has been most rapid in California, Arizona, and Utah. This increased manufacturing activity means boosts in plant investments and productivity.

Hercules, presently the valley's largest private employer with more than 4,000 employees, has Salt Lake roots dating back more than 75 years and is a prime example of adaptation to time's changing marketplace.

Hercules originally manufactured dynamite at its Magna, Utah, facility, which is named the Bacchus Works after the company's first general manager, T.W. Bacchus. In 1958 Hercules moved into the aerospace industry and secured a contract to build the third-stage rocket motor for the Minuteman missile. In 1962 the Bacchus Works plant employed more than 6,200 workers; however, declines in the defense industry caused employment to plummet to 1,700 workers by 1970.

Cold, winter weather produces large clouds of steam at the Standard Oil Refinery located in northern Salt Lake City. Photo by Steve Greenwood

The industry and the company rebounded during the 1970s, and Hercules secured contracts for the Pershing, Polaris, Poseidon, and Trident missiles. Hercules then expanded its product base to include graphite composites. The company's graphite composite material made it possible for the Voyager to complete its around-the-world, nonstop, non-refueled flight in December 1986. Much of this high-tech graphite airplane was built at Hercules' facilities in Magna, Utah.

In the mid-1980s Hercules broke ground on Bacchus West, located at the base of the Oquirrh Mountains. Bacchus West is the largest solid propulsion rocket motor manufacturing plant in the world. A $70-million Delta II contract followed, calling for Hercules to develop and produce the medium-launch vehicle's solid rocket motors for the U.S. Air Force. The plant is a fully-automated robotic manufacturing facility that was put into full operation in April 1987. Bacchus West employs approximately 165 people, making it efficient and competitive for winning government contracts against other propulsion companies.

Salt Lake is unquestionably becoming a center for space-related manufacturing. According to an "Aerospace Target Industry Study" conducted for Utah in 1986, "Utah provides an excellent location for aerospace manufacturing." What makes the area attractive to the aerospace industry are the educated, reliable work force, low levels of union representation, and vast areas of undeveloped land. The area's central western location also offers companies access to West Coast markets.

The new, 190,000-square-foot McDonnell Douglas Aircraft Company plant is evidence of that evaluation. Plant manager Richard Thomas says Salt Lake's aspirations to become an aerospace center are attainable. Along with the addition of McDonnell Douglas, Salt Lake has a well-rounded representation of firms that manufacture aircraft parts and systems. Montek Division of Dallas-based E-Systems Inc. produces flight control systems for military and commercial aircraft. Litton Guidance and Controls has manufactured navigation systems for military aircraft in Salt Lake for more than 30 years. Unisys, formerly Sperry, came to Salt Lake in 1957. The company manufactures computers for civilian and military applications. Several smaller firms also contribute to the area's manufacturing industry by making space parts and equipment. Many of these companies are here thanks to the state's Federal Procurement Office.

Since the program's inception in 1986, Utah's nine branches of the Federal Procurement Office have assisted small Utah firms in winning about 500 government contracts worth more than $50 million. "We've created or saved about 800 jobs," says James F. Odle, director of the program. "Our job is to make the process easier for prospective bidders." Funded by the state and the various Councils of Government throughout Utah, the program offers consultations with procurement experts, monthly workshops, a "bid

board," a network computer system, and ProcureSearch—an electronic subscription to *Commerce Business Daily,* which lists all federal contracts. "Currently, we have 600 companies actively using the ProcureSearch program," says Odle. "Where else can you find six million national contracts at your fingertips in a timely fashion, at no charge?"

When the governor created the program, Utah was one of only six states offering procurement assistance. Now, nearly every state with an economic development department also has a federal procurement office, patterning their programs after the Utah model.

The procurement program's success is evident from the number of contracts that have been awarded to Utah firms in the last 10 years. Since 1977 the Department of Defense has increased its prime contract awards to Utah firms from $227 million to $805 million in 1986. Today more than 400 Utah firms have contracts with the Department of Defense. Utah ranks 29th among the states in the dollar amount of its DOD contracts and 22nd on a per capita basis. The contracts pay more than $1,000 a year to each man, woman, and child in the state.

Union Park Center I is an example of the recent expansion of commercial space in the southern suburbs of the valley. Photo by Stephen R. Smith

MINING: "SOLID COPPER"

When Kennecott shut down its giant copper mine in 1985, people thought it was the end of an era, brought on by stiff foreign competition. Today the Bingham pit is a working mine again, with some 2,500 employees. Although this is a far cry from Kennecott's heyday when the company employed more than 7,000 workers, its $400-million plant investment is a sure sign of trust in the local economy. Kennecott modernized the plant by installing of an in-pit crusher, replacing the old rail ore-hauling system with a conveyor belt, and constructing a modern mill and concentrator. The modernization has increased efficiency and productivity, making the mine more competitive.

This Main Street crosswalk connects the ZCMI Center with the Crossroads Mall. Photo by Steve Greenwood

Below: Eagle Gate Plaza, one of Salt Lake's tallest structures with 22 floors, is one of the many new buildings that have changed the city's skyline in recent years. Photo by Stephen R. Smith

CONSTRUCTION: BUILDING NEW DIMENSIONS

The "clink of hammers" which Twain once heard echoing throughout the city is temporarily silent, waiting out the high vacancy rates. "Yet more has happened in the last 20 years to change Salt Lake's appearance than in the 100 years before that," says Fred S. Ball, president and general manager of the Salt Lake Area Chamber of Commerce. The building boom of new, class-A office buildings in the first half of the 1980s increased vacancy rates to more than 20 percent in the Salt Lake Valley, particularly in the downtown area. Many downtown firms moved south to the new structures springing up throughout the county. These firms were attracted by incentives such as free parking and convenient location.

However, after several years of decline, Salt Lake's office-leasing market is showing unmistakable signs of renewed strength, according to a recent edition of the annual "Salt Lake County Office Space Study," published by Wallace Associates. "We're cautiously optimistic about what we've seen," states Wallace's Howard Young. "However, we worry that optimism at this point could trigger an increase in development, which would likely turn the trends around again. There's still enough inventory out there to meet a few years of demand," he says. Salt Lake County's total supply of office space in 323 buildings reached 24.5 million feet in 1987.

Most of the older, ornate buildings that still stand, such as the landmark Romanesque City and County Building and the twin towers of Exchange Place, were built in the early 1900s. The Exchange Place towers, the Boston and Newhouse buildings, were once heralded as the Wall Street of the Rockies.

Between the 1930s and 1950s Salt Lake City's skyline changed very little. Most of the city's downtown buildings were erected during the first half of the 1980s, a direct result of the city's first master plan, the Second Century Plan, which architects and planners proposed in the 1960s. "All plans are dynamic, not static," says Dean L. Gustavson, one of the drafters of the document which resulted in the beautification of Main

Street, the Salt Palace, and the two downtown malls. Today the city has a renewed interest in updating that plan. An outside group of architectural consultants, the Regional Urban Design Assistance Team, was called in to study, evaluate, and recommend an urban planning project to enhance the city's natural surroundings and represent the spirit of the people.

The biggest challenge facing this team is the plight of Block 57 at the center of downtown. Most of the block, bordered by Main and State and 200 and 300 South, including the empty J.C. Penney building, is owned by one family. Mayor DePaulis, determined to get the project moving, established an advisory committee of community leaders to recommend a plan to develop the deteriorated block. The Salt Lake Redevelopment Agency has purchased the remaining property from individual owners, many of whom have had small shops there for years. The mayor is convinced they will attract a developer to this prime piece of property in the heart of town, but he says the development must be for the good of the community. This block has been suggested as a site for the proposed light-rail system's downtown terminal.

Salt Lake City's downtown comprises only an eight-block area. Although shops and businesses are within reasonable walking distance, everybody drives. Most pedestrian activity centers around the blocks where the two malls and Temple Square are located. The city's tallest building is The Church of Jesus Christ of Latter-day Saints headquarters, which stands 26 stories. One of downtown's newest additions is the lean, bronze, 22-story Eagle Gate, which was built snugly against the Beneficial Life Tower that rises 24 floors from the ZCMI Center mall. Another recent addition is the $600-million Triad Center, a magnificent example of contemporary urban design. The popular Utah Arts Festival is held on the Triad Center's expansive grounds every summer.

The Salt Lake International Center, an award-winning light industrial park west of Salt Lake International Airport, was constructed in 1975. The center includes two office complexes—Lakeside 1 and 2 on the shores of a small lake—the Airport Hilton, and several nationally known companies. The business park has surpassed everyone's expecta-

The famous Triad Center is seen here with its clock tower and flying flags during the spring season. Photo by Steve Greenwood

The Woodlands office complex occupies the site of a former drive-in movie theatre in the center of the valley. The building's eight stories added 135,000 square feet to the valley's supply of office space in 1985. Photo by Stephen R. Smith

tions, having transformed 900 acres of wasteland into a beautifully landscaped, award-winning business park.

On the other side of the central business district, two local developers in 1985 teamed up to create City Centre, a full-block project across from the Hall of Justice and the historic City and County Building. The first phase of the Price-Prowswood venture was the Salt Lake Area Chamber of Commerce building, a wide, 10-story cube encircled with balconies, brick walkways, planters, and rows of dark glass. Next to the Chamber of Commerce building is the state's Heber M. Wells office building, and another state-owned building will be added to the block in the near future.

Seeking more land and open spaces, developers have moved out to the suburbs of Sugarhouse, Murray, Midvale, and everywhere in between. Glass is in abundant supply, as evidenced in projects like the Forum, The Woodlands, Parkview Plaza, and Union Woods. These office projects have been successful because of their easy freeway access and central valley locations.

Several developers, such as Price-Prowswood and the Boyer Company, already have drawings and models for new projects sitting in their offices. The valley's next projects will likely include additional phases at City Centre and Block 57 downtown, and the Union Woods complex in the south valley. Planners envision Salt Lake City achieving a modern, elegant ambience, incorporating open spaces and Italianesque plazas similar to those in Portland, Seattle, and many California cities. Citizens, planners, and architects will work together to shape the future look of the city.

FINANCE AND RETAIL: AT YOUR SERVICE

Only two of Salt Lake's major financial institutions are locally owned: First Security Corporation, a long-standing family-owned institution committed to strengthening the arts and business climates of the community, and Zions Bancorporation, owned by the LDS Church. Other major Salt Lake financial institutions include First Interstate Bank, Key Bank, Valley Bank and Trust, and Continental Bank and Trust. All offer full service banking, including international banking services, and are FDIC insured with combined deposits of over $10 billion.

In addition, the Federal Reserve Bank of San Francisco in Salt Lake provides support services for the area's financial community. It expeditiously clears checks for businesses and financial institutions.

Utah's investment opportunities are many. The State's Division of Securities has tightened its regulation and enforcement policies, putting an end to Utah's reputation as a center for fraud. "Good opportunities are available for honest securities professionals and informed investors," says Governor Bangerter. "It's time to put the negatives behind us, invigorate local economies and do the best with what we have. It's time to invest in Utah." Reforms have turned Utah into a leader in state enforcement and investor education.

Venture capital sources are growing, too. The Utah Innovation Foundation each year sponsors the Utah Venture Capital Conference. This conference attracts between 30 and 40 people representing more than $2.5 billion in potential investment capital. This conference also gives Salt Lakers an opportunity to review new products, ranging from computer software to medical services. "Our response from the venture capital community has exceeded our expectations," says Brad Bertoch, executive director of the foundation.

Salt Lake City's retail industry has undergone dramatic changes as well. In the 1960s, when malls were built in downtown's north end near Temple Square, no one stopped anywhere else. Today most retail and pedestrian activity is concentrated in stores in and around the downtown malls. A new wave of strip malls and neighborhood centers, combined with existing regional malls in the suburbs, makes shopping convenient for all valley residents. Salt Lake's major department stores are ZCMI, Nordstrom, Weinstocks, Mervyn's, Sears, J.C. Penney, and Fred Meyers. Major grocery chains include Boise-based Albertsons and Salt Lake-owned Smiths and Skaggs Alpha Beta. Keen competition in the so-called "grocery wars" keeps prices moderate.

The Visitors Center at Temple Square is Salt Lake City's top tourist attraction. More than two million people visit the Temple grounds each year. Photo by Stephen R. Smith

TOURISM: A PLACE TO PLAY

Tourism is one of Salt Lake's major income sources, pumping $2 billion into the state's economy each year and generating $425 million in payroll for 47,000 Utahns and $110 million in state and local taxes. Salt Lake City's tourist trade supports businesses as varied as hotels, ski shops, and restaurants.

During the summer of 1987, city promoters saw definite signs of their marketing achievements: frequent traffic jams turned Bryce Canyon National Park into the Yellowstone of Utah, and inquiries at the visitors bureau at Salt Lake International Airport rocketed 231 percent.

Tourism officials say the area is fortunate to have a year-round tourist season. Eleven of Utah's 14 ski resorts are located within 60 minutes of the Salt Lake International Airport and attract winter visitors. In summer, thousands of tourists visit Utah's five national parks. Salt Lake City benefits from these year-round visitors; most arrive and depart through Utah's capital and spend some time there. Temple Square in the heart of Salt Lake City attracts 2.5-million visitors each year, more than Yellowstone National Park. Moderate gas prices and a weak dollar on the European market has kept many Americans home, where they have been rediscovering

Right: Canoeists paddle in the shadow of Mount Timpanogos east of Provo, some 40 miles south of Salt Lake City. Photo by Stephen R. Smith

Facing page: This intrepid skier tests his courage on Solitude's slopes. Photo by James W. Kay

Below: Arches of monumental proportions dwarf visiting hikers at Arches National Park, which is only a few hours away from the Salt Lake Valley. Photo by Mark Gibson

America—and Utah. Utahns, too, enjoy traveling their state, taking weekend excursions into the canyons, mountains, and deserts.

Although skiers comprise only a fraction of the 11 million visitors to Utah each year, the Salt Lake Convention and Visitors Bureau (CVB) concentrates its tourism efforts on marketing the "Ski Utah" product and group tours. Why? Because the skier spends more than $107 a day, whereas his or her summer counterpart spends only about $33. The ski resorts nearest Salt Lake—Snowbird, Alta, Brighton, Solitude, Park City, Park West, and Deer Valley—have become nationally known in the last 15 years.

Despite the fame of having the "Greatest Snow on Earth," Utah's promoters will have work to do to get the word out. An 18-month tourism study revealed that Utah and Salt Lake City do not have national reputations that attract visitors. Although people who have visited the area rate it highly, among the 21,000 Americans interviewed in the survey, only 10 percent named Utah as their first vacation choice. However, among respondents who had visited the area, 70 percent found it extremely to highly satisfying and 58 percent said they would recommend a Salt Lake vacation to friends. Most respondents rated the state high for its scenic beauty, its historical and cultural significance, and its recreational activities.

Tourism and convention officials attribute much of their industry's success to the private sector, saying that businesses are taking an active interest in promoting Utah by spreading the word to their associates.

Salt Lake targets most of its campaigns in Western states—Nevada, California, Arizona—and in the Pacific Northwest. At one time, Utah led the region in advertising dollars. Now other neighboring states have increased their budgets to compete.

The CVB spends most of its time and money on programs that are likely to produce the highest returns. Seventy percent of its annual $2.04 million budget is spent on attracting convention business whereas only 30 percent is spent on tourism programs such as skiing and group tours.

Since its reorganization in 1984, the Salt Lake Convention and Visitors Bureau has quadrupled convention sales. Prior to 1984 the CVB was a county government agency. Now it is a private organization staffed with marketing professionals who have experience in tourism and convention development. Eighty-five percent of the budget comes from a portion of the transient room tax; the other 15 percent is funded by member investment. In 1980 the CVB averaged 38,000 room nights a year, or about $2 million. It booked 101,000 room nights in 1985, 160,000 in 1986, and 164,000 in 1987. Rick Davis has big plans for Salt Lake's convention business in the next 10 years. "We're very

confident of our product," says Davis, who came to Salt Lake three years ago from Nashville where he doubled that city's convention business from $50 million to $100 million in seven years.

Why do businesspeople, clubs, and associations meet in the valley? For several basic reasons. Salt Lake's central western location and accessibility are attractive to planners because they usually guarantee a well-attended conference. The Salt Lake International Airport is only 10 minutes from town. The Salt Palace Convention Center ranks as one of the top 30 civic centers in the entire country. The center has 35 meeting rooms, a 12,400-seat arena, 5,000 parking spaces, and more than 200,000 square feet of exhibit space. Within walking distance of the convention center, several hotels offer 6,000 rooms with a view. "That's more than one could find in such proximity in Dallas, Denver, Phoenix or Los Angeles," says Davis. People can stay together and network without needing a shuttle service, which makes meetings in Salt Lake more productive and effective. First-class national hotels such as the Marriott, Embassy Suites, Holiday Inn, Hilton, Howard Johnson, Doubletree, Tri Arc, Red Lion, Residence Inn, and locally owned Little America have extraordinary service and are well maintained. The convention center is also only steps away from two indoor malls and Salt Lake's premier cultural showcases, the Capitol Theatre and Symphony Hall. Salt Lake's newest hotel, University Park, is located in Research Park on the University of Utah campus, not far from the Huntsman Special Events Center.

But what brings most people to Salt Lake City is its magnificent beauty and its close proximity to some of the most beautiful national parks in the country. Utah has five national parks (only Alaska has more) and five other parks are within a day's drive. Davis notes that the traditional, partying convention-goer is being replaced by people who bring their families with them and turn a meeting into a family vacation. And Salt Lake is well known for the attention it devotes to families.

Utahns are warm and friendly people. Service and hospitality come naturally to them. Salt Lake City is also a clean city where visitors can feel at home. The businessperson jetting from one city to the next will appreciate the accommodations, the friendly people, and the beautiful scenery. It is an ideal place to do business.

After a long and exhilarating day on the slopes, guests delight in the comfort and hospitality at picturesque Silver Lake Lodge in Deer Valley. Photo by James W. Kay

A study is now under way to find funding to enlarge the Salt Palace Convention Center and to create a hotel-type ballroom to replace the Grand Ballroom of the Westin Hotel Utah, which closed on August 31, 1987.

Like other cities, Salt Lake does have its share of problems. Sometimes the community tends to "stick its head in the sand" and pretend problems don't exist in "Happy Valley," says Mayor DePaulis. "The honest nature of the community has gotten us in trouble on occasion." Officials believe Salt Lake has learned and grown wiser from those mistakes.

Many Salt Lakers think there is no better place to live. Other residents dwell on the negatives, or are constantly apologizing for or defending the area's heritage and conservative way of life. These mixed reactions to the area are sending a confusing message to the rest of the country. Large-circulation newspapers in major metropolitan areas take this ammunition, load their guns, and take aim. Word-of-mouth is still the most effective form of advertising. It beats thousands of dollars in print and television advertising, and when it comes from friends or business associates, it is even more credible. People who live and work in Salt Lake City appreciate the things that set it apart from the rest of the country: clean streets, cosmopolitan atmosphere, wholesome life-style, and scenic beauty. Residents are justifiably proud of the city's acclaimed symphony and ballet, the healthy business climate, the high educational standards, and the emphasis on families.

Salt Lake's undefined image in the eyes of most Americans represents a unique marketing challenge. If leaders could do one thing to boost the economy, it would be to increase awareness of what the city has to offer. "We need to share Salt Lake's positives that we take for granted and increase the visibility of things we like about our homeland," says Rick Davis.

"Without worrying about appearing defensive, it's time for Salt Lake to stand up and take the offensive and debate the negative press point-to-point," says former mayor Ted Wilson, now director of the Hinckley Institute of Politics at the University of Utah.

Recently, Chamber President Fred S. Ball did just that. In response to a one-sided *Los Angeles Times* front page article on Salt Lake, Ball journeyed to L.A. and visited with the publisher and several key editors and writers. He responded to the article and invited the journalists to visit Utah and see the "true" city and area.

"Salt Lake has enormous wealth for a community our size," says Mayor Palmer DePaulis. "One of our most attractive assets is the unparalleled natural environment. If I wanted to go into business in today's competitive environment I'd be most concerned about my employees, that they were rooted in happiness. We are a global, cosmopolitan city not a provincial community. Everything we do must reflect that fact to compete in this new world, while preserving the best of what we have." An active Catholic and former student of the priesthood, DePaulis believes Salt Lake should not shed or deny its heritage but instead embrace its plurality.

The Mormon Church, too, has much at stake—with Salt Lake as its worldwide headquarters—in the area's economic well-being.

Community leaders, city planners, scientists, and engineers are also examining ways to improve air quality and transportation, crucial factors in Salt Lake's livability and, thus, economic vitality.

Business, government, education, and religious leaders all agree the city needs to focus its energy on economic development, but opinions differ on how to go about this. Jack Gallivan, publisher of the *Salt Lake Tribune* from 1960 to 1985, believes Salt Lake City will win its share of wealth. Many residents recall the boom times of the past and believe that only heavy manufacturing and mining will restore prosperity. Others see the answer in the clean, non-polluting industries of light manufacturing, high technology, finance, health services, and tourism. The valley's growing industries of the future will likely be aerospace, computer technology, telemarketing, food processing, medical research, and aircraft manufacturing joined by the basics of copper and steel.

History can provide a valuable lesson. Salt Lake must learn from its past and see it as a guiding beacon, not a warning signal. By expanding its industrial horizon, Salt Lake may discover the independent, peace-keeping wealth it previously only dreamed of.

BIONIC VALLEY

A writer in *BusinessWeek* magazine recently observed, "If there were really a Bionic Man, he probably would live in Salt Lake City. There he'd have ready access to an artificial heart as well as a broad range of spare parts, from prosthetic arms to artificial ears."

What Stanford is to Silicon Valley, the University of Utah is to Bionic Valley in Salt Lake City. The University of Utah and its Research Park, the fourth largest of its kind in the country, has placed Salt Lake at the forefront of America's most significant biomedical discoveries, technological advancements, and small company start-ups. Of the university's $65-million research budget, 90 percent is earmarked for the hard sciences and biomedical fields. *Science Digest* reported that "Salt Lake City is becoming Bionic Valley—the epicenter of a bioengineering effort that promises to shake up the entire health care system."

"The Salt Lake community is one of the most vibrant, exciting places in America," said a spokesman from the Stanford Research Institute, who had just completed work on a plan to attract high-tech businesses to Salt Lake City.

Salt Lake's mission to attract internationally known scientists such as Dr. Willem Kolff began in the 1960s. Former President James C. Fletcher (now in charge of the National Aeronautics & Space Administration) invited the Dutch-born physician Kolff to the university in hopes of transforming the school, the oldest state university west of the Missouri River, into a major medical center. Kolff came and others followed. His team of scientists and researchers brought about the marriage of the schools of medicine and engineering at Utah.

Collaborative research between the University of Utah schools of medicine, engineering, and computer science has produced bioengineering miracles such as the artificial heart, the Utah arm, INERAID (an artificial ear), the Wearable Artificial Kidney, and artificial blood vessels—all of which have been licensed to private firms for development and commercialization.

The University of Utah Research Park, one of the first and most successful of its kind in the country, was established in the late 1960s to provide a site for private research and development activities. The close proximity of the University of Utah's campus and health sciences center to high-tech companies is most likely the reason for the park's success. The University of Utah ranks 25th out of 2,000 American colleges and universities in funded research and is committed to introducing this research to the marketplace. Today research parks are springing up around the country, patterning themselves after the Utah and Stanford models.

The 320-acre park, with some 25 buildings housing 56 firms and nearly 4,500 employees, sits on the eastern foothills above downtown Salt Lake City between the University of Utah and Pioneer Trail State Park's historic "This Is the Place" monument. Research Park companies design and develop products based on new technology in medical science, engineering, biology, computer science, geology, and geophysics. Many of Research Park's companies, such as Evans & Sutherland, Terra Tek, Native Plants Inc. (NPI), and Northwest Pipeline, conduct more than $260-million worth of business annually and pay over one million dollars a year in property taxes.

One of the newest additions to Research Park, the University Park Hotel, opened in 1987. This 220-room hotel includes a modern conference center, a facility that park officials first requested in 1974. The prominence of the property's owner, the Kahler Corporation of

The University of Utah Health Sciences Center, world-famous for its artificial organ research and medical achievements, is located near the foothills five minutes east of downtown Salt Lake City. Photo by Stephen R. Smith

The 220-room University Park Hotel boasts many amenities, including an extensive conference center. Courtesy, the University of Utah Research Park

The 220-room University Park Hotel boasts many amenities, including an extensive conference center. Courtesy, the University of Utah Research Park

Minnesota (associated with the Mayo Clinic in Rochester, Minnesota), holds great promise for Salt Lake as the "Rochester of the West," believes Fred S. Ball, president of the Salt Lake Area Chamber of Commerce.

Careful planning and restrictive codes have created a quiet, park-like atmosphere with single-story buildings clustered around wide, open lawns and abundant parking lots. The park prescribes tough standards, prohibiting noise and noxious gas pollution. Companies housed in the park must conduct on-site research and development activities. Any assembly or processing must be related to these on-site research activities. "We want clean companies that pay high salaries," said the park's director soon after it opened. Many of these companies, such as Terra Tek and Evans & Sutherland, evolved from research initiated at the university.

The university owns all technology developed by researchers using its facilities, but it actively seeks to license the innovation to the inventor or private sources on an exclusive or nonexclusive basis. The Office of Technology Transfer makes this process of "academic capitalism" easier for university and private company officials.

"The University of Utah is committed to helping create and attract advanced technology industry. It accepts the responsibility to make the results of academic research benefit the public in the most efficient way," says James J. Brophy, vice-president of

Terra Tek is one of the many high-tech research firms located at the University of Utah Research Park. Here, an employee uses interactive graphics techniques for computerized fracture mapping of rock cores. Courtesy, the University of Utah Research Park

research at the University of Utah.

The university invests tens of millions of dollars annually in research covering a broad spectrum of technical areas, such as chemicals, agriculture, metals, energy, electronics, lasers, and medicine. The university also maintains a computerized technology data base, which systematically matches inventions to the technological needs of private industry.

"Our key function is to stimulate the faculty to recognize the benefits to themselves, to the university and to the community of disclosing their inventions to our office," says Norman Brown, director of the Office of Technology Transfer.

Yet rhetoric alone will not make Salt Lake City a center for high technology. Early on, state and city officials and educators realized that Salt Lake City would have to put its money where its mouth was if the area were to become a vital research center.

Dr. Thayne Robson, director of the University of Utah Bureau of Economic and Business Research, suggested a few years ago that the area should invest 5 percent, or $50 million, of state expenditures in creating a "hyper-competitive" environment for high technology.

Governor Bangerter agreed. Discovering new technologies, according to Governor Bangerter, "begins with research and extends in a continuous spectrum of development, test and evaluation, demonstration and, finally, commercialization." In Salt Lake City, economic development strategies are built upon these principles.

An example of the success of this strategy is the Centers of Excellence Program (COEP), established to promote targeted technologies developed at Utah's colleges and universities. With initial state funding of $2.5 million, more than 120 private companies and 14 federal agencies have invested over $38 million of direct support to Utah's 13 centers. More than 35 new companies have been created from the technologies developed at the centers.

The Centers of Excellence Program was patterned after the National Science Foundation's successful University-Industry Cooperative Research Centers Program, which combined government, private industry, and education resources. Tax dollars used to fund university research and to stimulate the transfer of that research into economic development are matched by industry at least two to one. The COEP has targeted seven technologies in which current research and development could be marketed: biomedical technologies, manufacturing technologies, engineering technologies, natural resources, communications and information technologies, space engineering and applications, and biotechnology.

The Biomedical Centers, all located at the University of Utah, have targeted four specific areas for research and development. The first area is developing biomedical devices such as the Total Artificial Hearts. Research at this center involves the design, fabrication, implantation, and monitoring of artificial organs, such as the famous Jarvik-7 artificial heart. The Center of Biopolymers at Interfaces, a joint effort of the colleges of engineering, pharmacy, and medicine, researches control of blood clotting and the development of artificial red cells and blood. The third Biomedical Center targeted for its research is the Laser Institute, which is internationally known for its development of new lasers used in medical and surgical applications. The Sensor Technology Center, the fourth Biomedical Center, is currently trying to attract new sensor research and development laboratories to Salt Lake.

Only one of the state's three Manufacturing Technology Centers is located in Salt Lake City: the Center for Materials and Advanced Manufacturing Technologies. This center studies properties, uniformity, reliability, and manufacturability of basic materials, giving scientists the competitive edge in aerospace, energy, semiconductors, communications, and medicine. And the Center for Engineering Design, located on the university campus, oversees the development of some 48 projects, including artificial limbs, drug delivery devices, artificial kidneys, and robotics. Currently, eight of the center's products are on the market, and six others are scheduled for release.

In the area of natural resources, two Centers of Excellence are located at the University of Utah. The Advanced Combustion Engineering Center is a joint venture between

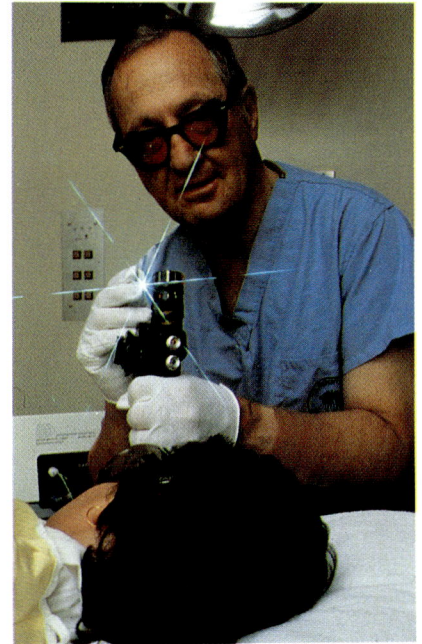

One of the four Biomedical Centers located at the University of Utah is the Laser Institute, which is a major force in developing new lasers with medical applications. Courtesy, Utah Centers of Excellence

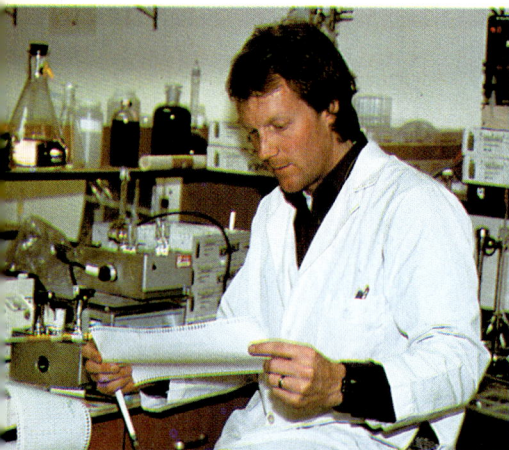

Above: Research and development is the main objective at the Center of Controlled Chemical Delivery. Courtesy, Utah Centers of Excellence

Right: The Advanced Combustion Engineering Center investigates new and more efficient ways to use low grade fuels. Courtesy, Utah Centers of Excellence

the University of Utah and Brigham Young University in Provo. This center develops computer systems that will use low-grade fuels cleanly and efficiently. The Center for Advanced Coal Technology is internationally known and hopes to expand the market for Utah's high-resin, high-volatile coals.

In communications, the university's Microelectronics Center develops materials for semiconductors and processes these materials into integrated circuits and new design tools. In addition, the Communication Research Center conducts basic and applied research through the university's electrical engineering and computer science departments.

Two space technology centers are located at Utah State University in Logan and Weber State College in Ogden. Both centers have attracted many new industries to the Wasatch Front in the past several years, such as Globesat, Incorporated, and IntraSpace. Globesat, in Logan, builds small satellites and launches them on board NASA's shuttle bays. The firm's technical team has experience in more than 400 rocket launches and 30 shuttle-related projects. IntraSpace, in Ogden, manufactures satellites for commercial use.

Associated with the Biotechnology Centers, the Center for Controlled Chemical Delivery is located at the university. This center researches and develops new techniques for administration of drugs to patients. For example, this center has patented a disposable, one-a-day patch for cardiac patients that administers the drug through the skin. All of

these research and development projects are part of the Utah Pioneer Partnership, a voluntary association of organizations within the state which support technological research.

Finding capital, taking risks, and learning to make crucial management decisions are essential to entrepreneurial success. The Utah Technology Finance Corporation (UTFC), another Utah Pioneer Partnership member, was created by the Utah legislature in 1983 to foster and encourage the development of emerging technology-based businesses. UTFC's primary function is to provide seed capital and support services to new entrepreneurs. This funding, up to $50,000, is provided through the Utah Small Business Innovation Program. Recipients are selected on the basis of their potential for high growth and job creation. The Utah Supreme Court upheld the statutory right of a state agency to channel public money to private start-up, high-tech firms. The justices ruled that organizations such as the UTFC are serving the public interest by promoting and funding "creative and innovative" technology-oriented firms.

Officials also realize that not all firms receiving funding will succeed, but "we can't afford to let fear or failure prevent us from helping promising entrepreneurs get off to a good start," says the governor.

Technovest, a new venture capital resource in the area, will pool funds for investment in the Wasatch Front counties. The Utah Technology Venture Fund 1 is designed

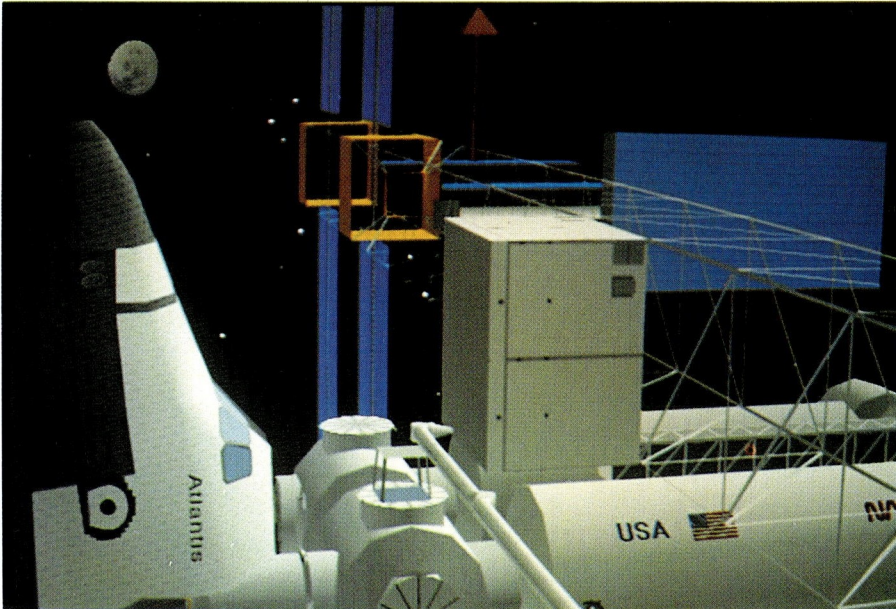

This computer-generated scene of the Space Shuttle Atlantis, *shown here docked with the planned MB-9 space station, was photographed directly from the screen of an Evans & Sutherland CT6 image generator. Courtesy, the University of Utah Research Park*

to alert larger venture concerns of promising Utah technologies.

The Utah Innovation Center (UICI) is a private firm that began as a National Science Foundation program at the university. Responsible for helping inventors get their products off the workbench and into the marketplace, UICI provides technical and business assistance, including work space. UICI is an incubator for ideas and projects that show promise as profitable, high-tech ventures. It takes an equity position in its client companies.

Out of these university programs have emerged some of America's most innovative companies. Some companies have become multimillion-dollar corporate giants. Other companies have yet to show a profit.

The founders of Evans & Sutherland, a world leader in graphic computer systems, conducted their initial research at the university. David C. Evans and Ivan E. Sutherland met in the 1960s while working on a computer consulting assignment. Evans, who received his doctorate in physics from Utah in 1952, was a professor at the University of California at Berkeley. Sutherland taught at Harvard. Evans developed computers that could be used by people who had no experience or training in computer systems. Sutherland was overseeing a Defense Department-funded program which was studying the feasibility of using computer-generated pictures for engineering design work. Together, they developed a graphic computer system that would be used for engineering, avionics, and projection systems. Today the company specializes in a computer graphic system that simulates real-world environments for use in pilot training, whole vehicle engineering, weapons-training systems, and avionics simulation. For example, NASA's Johnson Space Center in Houston, Texas, uses an Evans & Sutherland system for engineering design and astronaut training.

Evans & Sutherland also designs and builds sophisticated projection systems used in planetariums, including Salt Lake's Hansen Planetarium.

Founded in 1968, Evans & Sutherland today has more than 1,000 employees working at its Research Park site. The company has shown consistent growth and maintains an association with the university. The company still purchases time on the university computer, and several Evans & Sutherland computer scientists and engineers are adjunct faculty members of the University of Utah.

Native Plants Inc. (NPI) researches and germinates commercial quantities of Western American native plants. The plants, such as the Utah juniper tree, are then used for reclamation, landscaping, or erosion control along highways, dams, and mines. Using computer models, NPI conducts extensive laboratory tests on germination rates, calculates per-plant costs, and researches alternatives. The company also specializes in develop-

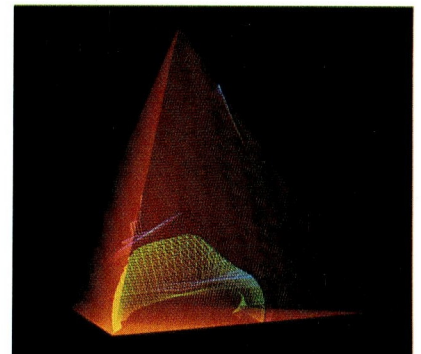

The Evans & Sutherland PS 330 graphics system was used to generate this image of a quaternary phase diagram, providing important information on the nature of chemical mixtures. Courtesy, the University of Utah Research Park

The Jarvik-7 artificial heart, a pioneering device in the heart transplant field, is manufactured by Symbion. Courtesy, Salt Lake Convention and Visitors Bureau

The Utah/MIT Dextrous Hand was designed at the University of Utah Center for Engineering Design, funded by the MIT Artificial Intelligence Laboratory. Licensed to the Salt Lake firm of Sarcos Inc., the Dextrous Hand is used as a device for machine dexterity research in the field of robotics. Courtesy, Utah Centers of Excellence

ing hardier, healthier agricultural crops. Staffed by scientists, NPI maintains a close relationship with the university's biology and horticulture departments and sponsors seminars and lectures. The firm's goal is to develop practical and cost-effective plants that will replenish the world's diminishing natural resources.

Terra Tek Incorporated, one of the park's first occupants, is a prime example of how one high-tech company can spiral into a series of spinoff companies. The firm specializes in geoscience research. "Terra Tek is really an incubator," says Sidney J. Green, president and chief executive officer. "We help create companies. As they mature, we push them out of the nest to fly on their own."

The firm originally grew out of federally funded research in the mechanical engineering department, which was studying the characteristics of various types of rock structures under extremely high pressure. Terra Tek's first contract, from the Department of Defense, assessed the impact of ground and air nuclear explosions on Minuteman ICBM silos. The Defense Department remains one of the company's major customers. Terra Tek also does work for large and small energy companies, as well as for aerospace firms such as Hercules and Morton Thiokol, Incorporated, in Box Elder County, Utah.

The Utah Arm is truly a modern miracle. Designed for agility and comfort, the elbow bends, the hand grips, and the wrist turns. Its advantage over previous prosthetic limbs is its ability to move in response to electronically processed muscle signals from the patient's remnant muscles. The arm is the brainchild of a group of physicians, prosthetists, and engineers specializing in control systems design, electronics, machine design, and computer technology. The arm was developed at the University of Utah and is manufactured in Salt Lake City by IOMED, formerly known as Motion Control. Motion Control was founded in 1974 by Dr. Stephen C. Jacobsen, director of the Center for Biomedical Design, and Dr. Willem J. Kolff, director of the Division of Artificial Organs. Jacobsen and his team of experts have invented two other biomedical devices—the phoresor, which administers medication through the patient's skin without the use of a needle, and the subcutaneous peritoneal access device, which is implanted in the abdomens of diabetics and enables insulin to be self-administered at a steady rate. The university retains the patent and other rights to any inventions. Motion Control has licensed the three patents and pays royalties on sales to the university.

Symbion, which manufactures the Jarvik-7 artificial heart, is another Salt Lake firm that owes its beginnings to University of Utah research. Symbion's Jarvik-7, used as a temporary heart replacement until a donor is found, has saved the lives of dozens of critically ill people at nine medical centers in five countries. Symbion is also credited with developing the INERAID cochlear implant. More than 100 totally deaf patients have received the implant. Patients use the tiny sound processor an average of 12 hours a day and report a sentence recognition rate of about 90 percent when they combine the tool with lipreading.

Today Utah has more than 50 firms involved in biomedical research and medical supply manufacturing, including large mass manufacturing firms like Deseret Medical—which has 2,000 employees—and small specialized companies engaged in "batch" production.

Since mid-1986 manufacturing in Salt Lake has added more jobs to the economy than has retail trade. The valley's high-tech firms are partially responsible for that increase, having jumped an annual six percent since 1980. According to Grubb & Ellis, a real estate services firm, the Salt Lake Valley's high technology growth rate topped those in most of the nation during that six-year period. The city is never listed, however, in national rankings of the top 15 technology centers because of its smaller high-tech base. Between 1980 and 1986 Salt Lake added between 5,000 and 6,000 new jobs to its high-tech foundation of 21,000 employees.

As with the rest of the nation, Salt Lake's business base is shifting from a goods-producing to a service-oriented economy. Service-producing jobs will continue to grow at a faster rate than goods-producing jobs. Although this new emphasis on high-tech development will require painful adjustments in some areas, it is the driving force behind productivity and competitiveness. Development programs, therefore, concentrate on restoring high-paying, high-tech jobs to the Salt Lake economy, offsetting the trend toward a service-only economy. High-tech industries are valuable not only because they improve the economy but because they raise the standard of efficiency and competitiveness of other industries that rely on high-tech equipment and research. Symbion, for example, anticipates rapid growth in heart transplantation and a corresponding increase in demand for its products. Since the first artificial heart implant in Barney Clark in 1982, the number of U.S. medical centers performing transplants has jumped from 10 to more than 80.

In the next few years, high-tech jobs in Salt Lake City will grow twice as fast as other industry jobs combined. The fastest-growing occupations in Salt Lake will be technical, service, and executive/administrative/managerial jobs.

A Terra Tek technician uses computerized axial tomography (CAT scan) to determine physical properties in the analysis of rock cores. Courtesy, the University of Utah Research Park

This transfer of talent and scientific knowledge from the classroom to the community will create a productive and diverse economy that will raise the standard of living, provide a clean working environment, and create higher-paying jobs for Salt Lake's educated work force.

However, Salt Lake City's prime products in coming years will be inventions and ideas. To succeed in this new economic environment, the community will have to devote itself to attracting new technology and bright people and marketing these resources.

University researchers say there are plenty of great ideas just waiting for the right person or company to take an interest. Salt Lake's "Bionic Valley will be Utah's hottest growth industry for years to come," says *BusinessWeek.*

The foundation for high-tech industry appears to be in place. The university's artificial heart program has just received $6.7 million from the National Institute of Health to develop a totally implantable artificial heart, one of the largest federal contracts the University of Utah has ever received. "This is a very grandiose and a very large undertaking . . . but some of the technology indicates this is possible," says Dr. Don B. Olsen, director of the Artificial Organs Division of the Institute for Biomedical Engineering. Utah's program is now competing with three other programs to retain its funding for artificial heart research and technology. The Utah 100 artificial heart is much smaller than the other totally implantable devices and is designed to fit in a small chest. Officials predict the first human implant of the Utah 100 is at least eight years away. Dr. Olsen believes natural heart transplants could save between 35,000 and 50,000 lives each year if there were enough heart donors; however, at present only 5,000 donor organs are available.

Salt Lake is also a center for genetic research. Although still in the planning stages, the $50-million George and Dolores Eccles Institute of Human Genetics (funded through the Howard Hughes Medical Institute and the Eccles Foundation) on the University of Utah campus is predicted to be the nation's leader in genetic disease research.

"We have moved into a high-tech, high-frontier world without losing either our roots or our values. Hard work, honest labor, optimism, strong family ties, a commitment to education, and a strong belief in the free enterprise system create a positive climate for business to thrive," says Governor Bangerter.

CROSSROADS OF THE WEST

Visitors who fly into Salt Lake City for the first time are often amazed at the grandeur of the scene below. From the east, mountain ranges rise majestically as if from a lunar landscape. Below lie the Rocky Mountains, the Uinta Mountains—the tallest mountain range in North America to run east and west—and the Wasatch range. Their summits and valleys seem to go on forever, yet their snow-covered slopes soon disappear beneath cottonball clouds. As the plane approaches the valley, Lone Peak and Mount Olympus seem to step aside, revealing the Salt Lake Valley beyond their granite shoulders.

From the west the expansive desert scene is as different from the rugged mountain ranges as the moon's terrain is from earth's. Dusty sagebrush, red sunburned rocks, and a brown-colored basin create a colorful landscape leading to the chalky shores of the Great Salt Lake. Sprawling across the open spaces of this desert basin are Tooele Army Depot, Dugway Proving Ground, and acres of Air Force testing ranges. Occasional mountain ranges wrinkle the western desert, sometimes rising 12,000 feet from the desert floor. These craggy ranges—the Deep Creeks, the Newfoundlands, the Oquirrh, and others—seem like tiny oases in the Great Basin. While no spot on earth is truly remote in today's world of superhighways and jumbo jets, this region is mostly pristine wilderness, a refuge for hundreds of species of endangered or threatened plants and animals. The Deep Creeks is the only mountain range in the interior of the Great Basin with an abundance of water. It is home to one of earth's oldest living trees, the bristlecone pine. Pronghorn, bison, mule deer, desert bighorn sheep, and wild horses roam the canyons, valleys, and slopes of these ranges. Golden and bald eagles soar overhead, and the endangered peregrine falcon nests nearby. The wetlands of the Fish Springs National Wildlife Refuge shelter swans, geese, ducks, and other birds in the middle of this sun-scorched ecosystem.

The Oquirrh range forms the western side of the basin known as the Salt Lake Valley. Here lies the world's largest open-pit copper mine, plainly visible from any point in the valley.

Even that famous nineteenth-century traveler and philosopher, Mark Twain, recognized the wild side of traveling through this part of the country:

It was pleasant . . . to reflect that this was not an obscure, back-country desert, but a very celebrated one . . . All this was very well and very comfortable and satisfactory—but now we were to cross a desert in daylight. This was fine—novel—romantic—dramatically adventurous—this, indeed, was worth living for, worth travelling for! We would write home all about it.

Today this area is much the same as Twain found it over a century ago, except that now, instead of a stagecoach watering station, visitors will find the Salt Lake International Airport, only ten minutes west of Salt Lake City's central business and financial district.

If rankings are important to air travelers, Salt Lake International Airport (SLIA) is the 25th busiest in the country and the 35th largest in the world. A four-month security test of 28 airports around the nation found SLIA to be the third best in security checks. According to a Federal Aviation Administration (FAA) study, when testers attempted to pass through security check stations carrying guns, defused grenades,

The nation's longest interstate, I-80, runs east and west through the Salt Lake Valley, while Interstate 15 extends north and south. Photo by Stephen R. Smith

Wild ducks take flight over the frozen marshland northwest of Salt Lake City, which is visible in the distance. Salt Lake Valley residents benefit from the beauty and abundance of wildlife the area has to offer and take great care in preserving this natural state. Photo by Steve Greenwood

A Delta Air Lines jet takes off over the main terminal of the Salt Lake International Airport, which is only minutes away from Salt Lake's central business district. Photo by Steve Greenwood

and other mock weapons, SLIA security officers detected 92 percent of these objects.

Salt Lake International has the highest percentage of on-time arrivals and departures of the nation's 27 largest airports. This figure is based on statistics compiled by the United States Department of Transportation.

From the start, Salt Lake International was one of the world's foremost aviation centers. In 1920 Salt Lake City purchased 100 acres surrounding a narrow, rutted landing strip for $40 an acre. Known then as Woodward Field, this airstrip accommodated airmail carriers. But Salt Lake's aviation history really began in 1926 when two Salt Lakers donned leather helmets, goggles, and parachutes and climbed into the open cockpit of a Douglas M-2 biplane. Their Western Air Express flight took off from Woodward Field and landed eight hours later in Los Angeles. These two men, scrunched between bulky U.S. mail sacks, were the first commercial air passengers in the United States.

In each decade since the 1920s, Salt Lake International has been expanded and remodeled. Only the outbreak of World War II disrupted the airport's progress as a commercial air center. During the war the U.S. Air Force used the facility as a training base and replacement depot. To accommodate the trainees, the city built Salt Lake City Municipal Airport II in the southwestern section of the valley. The jet age of the 1950s spurred a new era of development at the airport. To accommodate the increasing air traffic, three runways were added. Today "Salt Lake International can handle anything that flies," says Lou Miller, airport director.

The 1960s were a time of whirlwind planning and policy-making activities. Anticipating the phenomenal growth in the airline industry and in the Salt Lake Valley, airport and city officials initiated a master plan to study and carry out an airport expansion plan. The Airport Master Plan was completed during the summer of 1988. Forecasters predict that the number of travelers from Salt Lake City International Airport will more than double by the year 2006.

The most significant change at the airport came in 1982 when Western Airlines established its domestic hub in Salt Lake City, expanding the airport's operations threefold. Between 1982 and 1986 Salt Lake International was the nation's fastest-growing airport. "The hub has changed the entire character of the airport," Miller says. As a major hub for connecting flights, the airport has had to accommodate more concourses and gates.

Passenger traffic through Salt Lake International rose from 4 million in 1981 to 6.5 million in 1982. On April 1, 1987, the 60-year-old Western (the nation's oldest continuously operating airline) merged with Delta Air Lines. Since this merger, Delta Air Lines has increased Salt Lake City's status as a transportation hub. In 1987 Delta added a marketing and reservations center—which employs 500 people—and an aircraft maintenance hangar at an estimated cost of $25 million. Now the nation's fourth largest air carrier, Delta Air Lines employs 30,000 workers systemwide, with 3,000 employees in Salt Lake City. Overall, Delta Air Lines contributes more than $200 million a year to Salt Lake's economy. "Western's Salt Lake hub was the jewel in the crown that prompted Delta to merge with Western," says Bill Jackson, Delta Air Lines manager of public relations.

"Delta's expansion at Salt Lake City confirms the importance of this city and, in fact, this state and region to our systemwide operations in the U.S. and overseas," says R.W. Allen, Delta Air Lines president and chief operating officer.

Just as the railroad was responsible for putting Salt Lake on the map in the nineteenth century, Salt Lake International Airport is now jetting Salt Lake City into the twenty-first century. "Since it was first settled, Salt Lake has been the crossroads of the West," continues Jackson. "Delta needed a city with a good overall economic climate, a central location and an excellent working relationship with the government and the airport board. Salt Lake has those things," Jackson explains.

In addition to Delta Air Lines' new hub center, the airport has also added a $2-million Federal Express sort station and a new U.S. Postal Service mail facility. In 1987 McDonnell Douglas Aircraft opened a new manufacturing plant at the northwest end of the airport. The plant was intended for assembly work on C17 cargo planes for the U.S. Air Force, a costly project that, in light of government's deficit problems, has been postponed. Instead, employees at the plant perform subassembly work on MD80 commercial planes.

Nine commercial and three regional commuter carriers serve the Salt Lake area with more than 300 flights a day. Salt Lake International has nonstop flights to 67 cities and same-plane service to 105 cities. "We have easy connections to the entire world because we have nonstop service to all hub cities in the country," says Miller. Delta Air Lines handles 70 percent of the airport's total traffic, providing service to 50 percent of

Above: Delta Air Lines established its western hub in Salt Lake in 1987 after merging with Western Airlines. Photo by Stephen R. Smith

Top: Delta operates a marketing and reservations center at the Salt Lake International Airport and is one of the valley's leading employers. Photo by Stephen R. Smith

A Delta Airplane takes off into the Salt Lake sky. Photo by Steve Greenwood

all passengers traveling to or from Salt Lake City.

In fact, all the airlines are contributing to Salt Lake City's growing reputation as a vital and prospering economic center. Eastern's Salt Lake reservations center employs more than 400 workers. SkyWest Airlines, Utah's regional commuter carrier, was recently named "Regional Commuter Carrier of the Year" by the industry's leading magazine, *Air Transport World*. This annual competition included hundreds of regional carriers serving the nation from Alaska to California and Florida to Maine. Based in St. George in southern Utah, the 17-year-old company now has more than 1,000 employees throughout the Southwest. Delta Air Lines recently invested $6 million to acquire approximately 20 percent of SkyWest Airlines' common stock.

Indeed, Salt Lake International Airport has become a key element in the economic growth of Utah, bringing businesspeople, tourists, and skiers into the Salt Lake Valley. The airport also provides jobs to more than 6,400 employees (an 18 percent increase since 1985) and pumps more than $400 million annually into Utah's economy.

Traffic through Salt Lake International Airport is expected to double in the next 13 years, from 11 to 22 million passengers annually. Airport officials also anticipate a 129 percent jump in the number of passengers boarding aircraft, including stopovers and transfers. Commercial aircraft arrivals and departures will increase 140 percent, while air cargo will jump 110 percent. In 1985 Salt Lake International handled 85 departures and arrivals during peak hours, but officials project 100 operations per peak hour in 1991, 111 in 1996, and 133 by 2006. Annual aircraft movements will total 424,400, up from about 275,000 in 1987. Like other metropolitan centers around the country, Salt Lake International must be prepared to accommodate this tremendous increase in passengers.

To increase airport capacity and efficiency, airport officials have proposed a master plan to the FAA. This plan calls for adding several high-speed taxiways and a runway (pending environmental wetlands issues since much of the surrounding land is home to thousands of migrating Canada geese, Great Blue herons, and other species of birds) and installing an instrument landing system for air traffic arriving from the north.

Owned by Salt Lake City and operated by the Salt Lake Airport Authority, Salt Lake International Airport's 7,500 acres include enough space for future expansion plans. By contrast, New York City's John F. Kennedy Airport occupies about 5,000 acres. Facility expansion averages about $22 million annually in expenditures. Award-winning landscaping, including fountains, ponds, shrubbery, and signage, has been completed in the last few years.

In the next five years the airport plans to expand baggage claim and ticketing areas in Terminal Two and add a four-level parking structure. Also planned is an elevated, over-the-curb walkway that will connect the garage with the terminal. Another 10-year plan calls for adding gates to Terminal Two, and installing by the year 2006 a shuttle system that will carry Delta Air Lines passengers to their connecting planes. By the year 2006 planners want to increase the number of gates from 48 to 76.

Planners are also carefully trying to avoid the expansion obstacles facing neighboring Stapleton Airport in Denver. Noise over neighborhoods surrounding Salt Lake's airport forced the city to spend $6 million to purchase 57 new homes immediately south of the airport. "All municipalities and jurisdictions must protect the airport's ability to operate," says Miller. "Encroachment can't stifle our growth, or we'll be forced to lose our investment, cut flights or start all over again." Denver's problems haven't gone unnoticed in Salt Lake. Prior to its merger with Delta, Western Airlines used the opportunity to tout Salt Lake's "easy in, easy out" in its print advertisements.

"Salt Lake has no built-in delays like other major cities," says Miller. In fact, it rarely closes due to inclement weather. Snow has forced the Salt Lake International Airport to close only once in the last nine years, although heavy snowstorms occasionally slow operations. Salt Lake International's snow-removal crews have won national awards for their efficiency in seven of the last nine years.

The modern, efficient airport is not the only evidence of Salt Lake's reputation as a transportation hub. One hundred and thirty years ago, workers for the Union Pacific

Flags of many nations fly over the main circle of the Salt Lake International Center. The center, located west of the Salt Lake International Airport, is a large, successful industrial park. Photo by Steve Greenwood

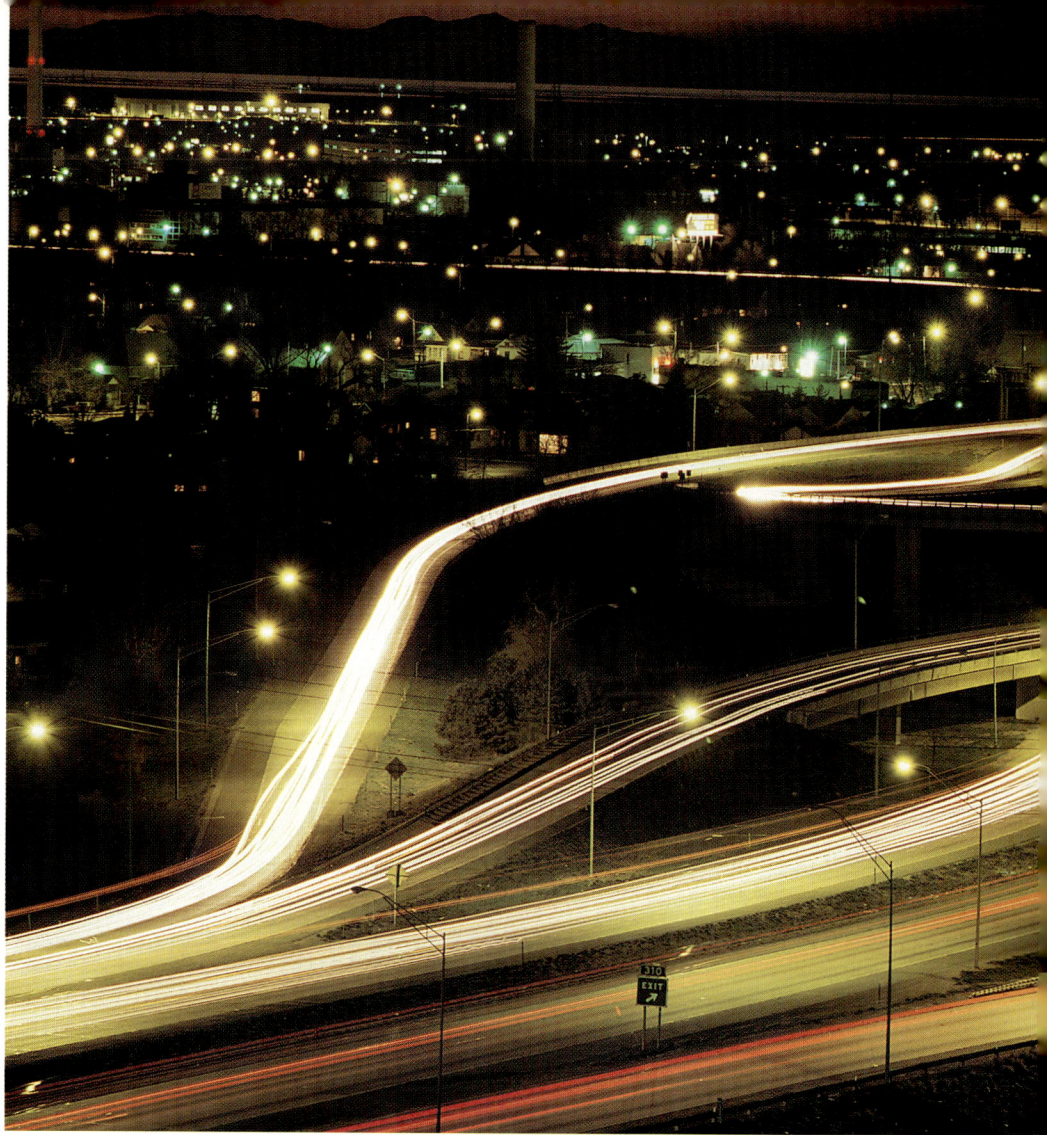

Interstate 15 is the major north-south freeway through the Salt Lake Valley and it is pictured here looking north at the 6th South off-ramp. 6th South is the main exit into downtown Salt Lake City. Photo by Steve Greenwood

and Central Pacific railroads drove the golden spike connecting the eastern and western railroads at Promontory Point in northern Utah. History does seem to repeat itself. In 1987 construction workers completed Interstate 80, just west of Salt Lake, adjacent to the airport and the Salt Lake International Center business park. This was the final link in the east-west continental interstate system.

The Salt Lake International Center, built in 1975, transformed 900 acres of barren flatlands west of Salt Lake International Airport into a green, park-like business center. Two glass office complexes, Lakeside I and II, sit near a large duck pond. Flag Plaza is the heart of the center. Mature pines and colorful flower beds line the park's wide boulevards, which are named for famous aviators: Charles Lindbergh Drive, Wright Brothers Drive, and Wiley Post Way. The park contains some of Salt Lake's prime industrial space and high-tech companies, such as Beehive International, Skaggs Telecommunications, and other prestigious national firms. The center offers excellent air, rail, and truck transportation service. Salt Lake International Airport is immediately east of the park. Union Pacific and Denver Rio Grande Western provide national rail service. The newly completed Interstate 80 provides freeway access to the center.

Salt Lake is an ideal distribution center. The Intermountain Foreign Trade Zone, located in the northeast section of the Salt Lake International Center, is the first such zone to be established in the Intermountain West. A foreign trade zone is an area considered outside the customs territory of the United States, and therefore exempt from custom tax and duty payments, customs procedures, and storage quota restrictions. A foreign trade zone also provides manufacturing, distribution, and warehouse space. Companies may import a product into the United States, put it into the foreign trade zone, and not pay duty taxes on it until the product is withdrawn.

Utah's Freeport Laws make the state an ideal location for warehousing. The state has no ad valorem taxes on inventory of any kind. An exporter can centralize distribution and easily transport goods to seven major West Coast ports. More than 1,500 freight carriers operate throughout Utah, and some 40 major interstate carriers are based here. Interstate 80 and Interstate 15 converge in the heart of the valley, giving com-

A Union Pacific freight train pulls out of Salt Lake City and heads north through Davis County. Photo by Steve Greenwood

panies easy access to other major markets.

The nation's largest distribution center, the Freeport Center, is located just north of Salt Lake City in Clearfield. The owners converted the former Clearfield Naval Supply Depot, built during World War II, into warehousing space in the 1960s. Today the Freeport Center has more than 7 million square feet of low-cost manufacturing and distribution space. Located 750 miles inland, the center enables shippers to deliver goods to any Western city within two to three days. Merchandise also can be shipped directly to Freeport Center, stored for any length of time—exempt from inventory taxes—and then re-shipped to its final destination at a continuation of the through-rate.

The center also has its own container and piggyback trailer handling facility for shipments to Hawaii, Alaska, or foreign ports. There is no additional cost to customers without railsidings or overseas customers to substitute piggyback trailers or containers on the outbound move. U.S. Customs service is available on a daily basis as well.

With a single inventory housed at Utah's Freeport Center, companies save money and increase profits by reducing total inventory costs, increasing turnover, reducing stock outs, and eliminating interwarehouse transfers. Companies such as Fisher-Price Toys, Fram Corporation, Clover Club Foods, General Mills, and Wilson Sporting Goods have realized these benefits.

Many food processing companies—such as All American Gourmet in the Freeport Center, Southland Corporation in Salt Lake City, and Stouffer's in Springville—also benefit from Salt Lake's central location. It is easy to get raw materials here and easy to ship the finished product out. Other major industrial and business parks in the Salt Lake Valley include Centennial Park (a subsidiary of Union Pacific Corporation), Technology Park, Wagner, Decker Lake, Metro, and Time Square.

Utility rates in the Salt Lake area are relatively low compared to the rest of the nation. Mountain Fuel, based in Salt Lake, is one of the nation's leading multifaceted regional energy companies and offers some of the lowest rates in areas of Utah and Wyoming. Mountain Fuel and its affiliated companies, owned by Questar Corporation, are involved in the exploration, production, transportation, underground storage, and distribution of natural gas.

Electricity to most of the valley is provided by Utah Power and Light Company, a subsidiary of PacifiCorp. Bountiful City and Murray City have their own municipal power service.

AROUND TOWN

Getting in and out of Salt Lake City is a breeze, but if current projections are correct, by 2010 there will be almost 50 percent more cars on Interstate 15. Engineers and planners who have studied transportation along Utah's metropolitan strip say traffic congestion could be reduced in the future by adding more lanes to Interstate 15, adding more Utah Transit Authority (UTA) buses, and by introducing a light-rail system. A light-rail system would be able to remove about 12,000 cars and trucks from the busy I-15 corridor. UTA is exploring the possibilities of building a light-rail system from the southern end of the valley to the downtown area.

Salt Lake already has one of the best public-transit systems in the country. The UTA, Salt Lake's mass-transit operator, has received the most prestigious award for bus service in the United States, the Public Transportation System Outstanding Achievement Award. Compared with other bus systems, which typically spend from $4 to $6 per revenue mile, UTA spends $2.24 per revenue mile. UTA also provides service to 70 percent

Below: When construction on this portion of Interstate 215 is finished, Salt Lake Valley's belt route will be complete. Photo by Stephen R. Smith

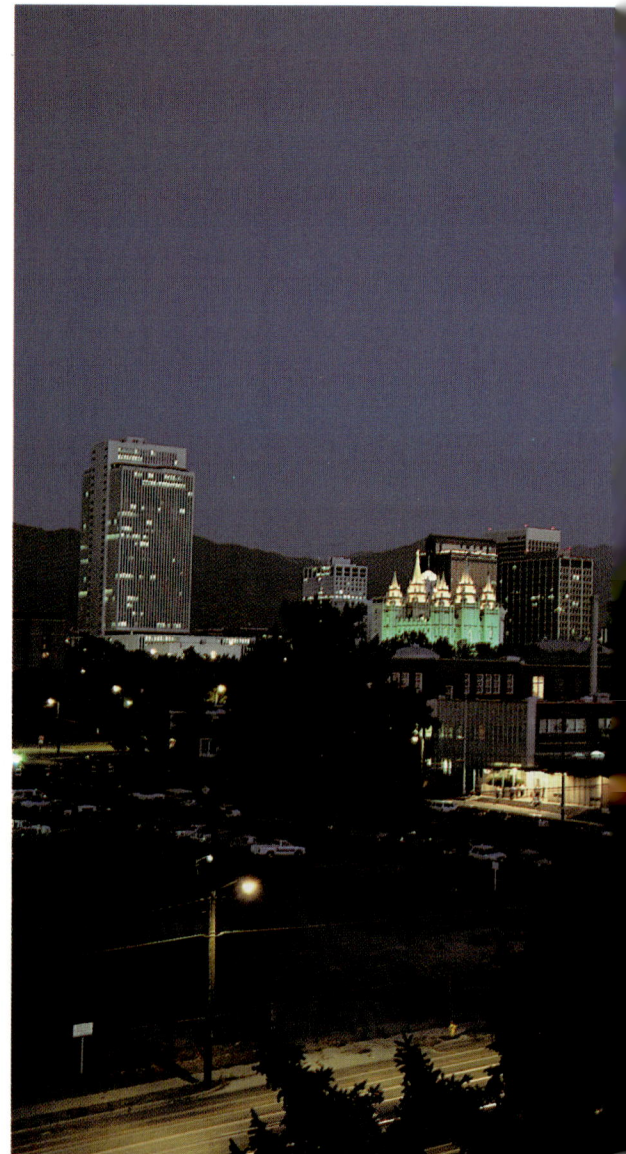

of the population in Salt Lake, Davis, Weber, and Utah counties, with bus service within one-quarter mile of residents' homes. UTA buses arrive within five minutes of the scheduled time 93 percent of the time. In the last few years UTA has expanded and improved its service and continues to be responsive to the needs of the community. Mass transit reduces air pollution, congestion, and highway use; conserves energy and natural resources; and provides transportation to low-income and disabled residents.

A new system called Flextrans (flexible transportation service) was implemented in 1988 to accommodate disabled passengers who could not get to bus stops. Flextrans' curb-to-curb public transit service picks up people at their door and takes them to the nearest sheltered bus stop. Here, they may take advantage of lift-equipped bus service.

UTA's 400 transit vehicles follow more than 100 daily routes in an 1,800-square-mile area populated by more than one million people. The base fare is 50 cents for peak hours. Handicapped and senior citizens pay 25 cents. Contrasted to Hertz Corporation's estimate of 73 cents per mile to own and run a car, the bus is an extremely cost-effective means to get around town. In 1986 total savings for all UTA riders was $8.8 million, and this does not include the benefits of reduced driving stress, on-bus leisure time, and increased safety.

UTA also operates special trolleys between downtown hotels, office buildings, and shopping malls.

Busy Interstate 15 conveniently connects Salt Lake with other major markets. Photo by Stephen R. Smith

GETTING BACK ON TRACK

Electric trolleys, the first public transit system in Salt Lake City, were introduced in 1889. A century later, Salt Lakers are again considering the benefits of letting someone else do the driving. The Federal Urban Mass Transit Authority evaluates time saved in commuting to work at four dollars an hour nationwide. If that figure holds true for Utah commuters, they would gain a total of $2.6-million worth of time each year by using mass transit.

Cities such as San Francisco, Pittsburgh, New Orleans, Philadelphia, and Boston have relied on mass transit for almost a century. Other cities that abandoned their trolley systems to make way for freeways—such as Buffalo, Sacramento, San Diego, and Portland—are now introducing new light-rail transit systems (LRT) with great success.

LRT operates on new tracks laid in the streets or existing tracks abandoned by a railroad. LRT cars are electrically powered by overhead wires.

An LRT system has been proposed for an area running from 106th South to downtown Salt Lake City. It could be in service by 1994 at the earliest. Stations along the route would serve Sandy, Midvale, Murray, South Salt Lake, and Salt Lake City. Seventy-two percent of Salt Lake Valley businesses are located within this area. The downtown terminus could be located underground below the blighted Block 57, which is already scheduled for major redevelopment.

To install the system by 1994, federal, state, and local approvals must be obtained. "The public itself has to decide this is what they want to do and what they're willing to pay for," says John English, UTA operations director.

Educating the public on the benefits of mass transit will be the supporters' biggest challenge. Planners hope to have the LRT in place before construction begins on the freeway improvements. This will offer commuters an alternative during construction of the freeway and introduce LRT to a greater number of potential riders.

Salt Lake's LRT could operate on the existing Union Pacific right-of-way and could be funded by a local option gasoline tax or expansion of the sales tax to include gasoline. Officials believe this would be fair because the tax is aimed at those who create traffic problems.

THE WORD IS OUT

Highways, railways, and flight paths are not the only links connecting Salt Lake to the rest of the world. Salt Lake is also the largest geographical media market in the country and the center of communications, news, entertainment, and advertising for the Intermountain West. Salt Lake's television signals transmit throughout all of Utah, the western half of Wyoming, southern Idaho, and portions of Colorado, Montana, and Nevada. Visitors are often impressed with the quality and professionalism of Salt Lake's three network television affiliates. Salt Lake was actually among the first cities in the country to have a television station, an important claim to fame since Philo T. Farnsworth, known as the "father of television," came from Utah.

KTVX Television, Channel 4 (Salt Lake's ABC affiliate) signed on the air in 1948 as Utah's first and the nation's 13th television station. Salt Lake's W6XIS, as KTVX was then known, was the only station between St. Louis and the Pacific Coast. W6XIS was broadcasting before stations in San Francisco, Seattle, Portland, or San Diego. KTVX built the nation's third highest transmitter atop Mt. Nelson in the Oquirrh Mountains in 1980. Owned by United Television of Minneapolis, KTVX has recently devoted considerable time and resources to improving its coverage of local news by conducting live televised interviews with local and national newsmakers. This effort has significantly increased KTVX's ratings. To capture a new audience for its evening news, KTVX broadcasts its news program at 5:30 rather than 6:00 p.m.

The Mormon Church is largely responsible for the sophistication of Salt Lake's communication networks. Bonneville International, owned by the Church, is the parent company of KSL Television, Channel 5 (Salt Lake's CBS affiliate), KSL Radio, and several state-of-the-art communications, satellite, and production systems. The Mormon Tabernacle Choir's weekly radio show has been broadcast on KSL Radio throughout the world every Sunday morning since the 1920s, the longest-running broadcast in history. KSL leads the pack in viewer ratings with its nightly "Eyewitness News." KSL recently moved its station to the deluxe Triad Center. *Electronic Media,* an industry trade journal, called the new station "the most spectacular radio and television facility in the United States." According to rating sources, KSL's television news is the most watched local news program in the country. Its weeknight news anchor team has worked together longer than any other team in the nation.

KUTV Television, Channel 2, is Salt Lake's NBC affiliate and usually runs neck-and-neck with KSL for viewer ratings. KUTV attracts a younger audience than does KSL and is best known for its "Project 2000" series. "Project 2000" is a diverse group of Utah citizens, including a local newspaper publisher, attorneys, legislators, teachers, a Utah Supreme Court justice, and a representative of the League of Women Voters, that meets to discuss ideas for the KUTV series and to look at "what the state ought to be in 20 years."

KSTU Channel 13, Utah's strongest independent station, runs old movies and other family favorites but does not produce a news program. It offers viewers an alternative to network programs.

Educational programming on KUED, Channel 7, and KBYU, Channel 11, offers children's shows, science and nature presentations, documentaries, and theatrical performances.

Dozens of AM and FM radio stations offer various music and program formats, from rock and roll and jazz to country-western and news.

Two daily newspapers serve the Salt Lake metropolitan area. The morning *Salt Lake Tribune* has the largest circulation, and the *Deseret News* is delivered in the afternoons. The Newspaper Agency Corporation handles circulation, advertising, and production for both papers. Both more than 100 years old, the newspapers have recorded the events and issues that have shaped Utahns' lives.

Major area magazines include *Utah Holiday,* which appeals to a wide audience of upscale, professional readers. It covers fine arts, culture, and business trends. Its theater, restaurant, and arts guides give helpful suggestions to visitors as well. *The Event* is a monthly tabloid containing information on the arts and theater along the Wasatch Front. *Wasatch Sports Guide* is a monthly publication for Utah's outdoor enthusiasts and covers activities like bicycling, skiing, mountain climbing, backpacking, and river-running. *Network* is a publication written for women "and the men with whom they live and work." It contains articles about relationships, career development, jobs, business, job politics, families, and Utah culture. *Catalyst* is a publication for anyone interested in health, nutrition, nature, and metaphysics.

Salt Lake's diverse media market is testimony to the city's cosmopolitan atmosphere. At the crossroads, Salt Lake has something for everyone.

Salt Lake is the largest geographic media market in the country. Three network-affiliate broadcast stations reach all of Utah as well as parts of Idaho, Wyoming, Colorado, and Nevada. Photo by Stephen R. Smith

CHAPTER SIX

HEALTHY, WEALTHY, AND WISE

Education is one of the foundations on which people build a community, maintain their quality of life, and preserve their heritage. "Today's schools still fulfill this role, nourishing the fundamental, conservative values that have contributed so much to our quality of life: respect for hard work, for self-reliance and dependability, and for a healthy lifestyle and high moral standards; and respect for the arts as a basic component of education that fosters self-discipline, creativity and an appreciation of beauty," reads a Utah State Board of Education report.

As the twenty-first century approaches, Utahns will face the same challenge confronting other states: how to maintain quality education while keeping expenditures low. Reform is inevitable, and Utah is among those states already setting new education standards.

Utahns are some of the most literate people in the country. According to the United States Bureau of Census, Utah's literacy rate is 94 percent compared to the national literacy rate of 87 percent. Most new businesses that have moved to Salt Lake City—American Express, Delta Air Lines, Fidelity Investments, McDonnell Douglas—say this human resource was a major selling point.

Education is a priority on Utah's agenda, and so it should be. After all, 25 percent of the state's population is between the school-age years of 5 and 17, compared with 18.7 percent in the United States, and another 10 percent is between the ages of 0 and 4.

Utah has long been known as the baby capital of the country, where preschools and day-care centers outnumber even toy stores. Between 1980 and 1987 Utah's school-age population increased nearly 23 percent. During this same period, Colorado's school-age population remained fairly even and Massachusetts' actually declined 15 percent. Only Alaska, with an increase of 20.7 percent, comes close to matching Utah's school-age population. Although Utah's total fertility rate (a measure of average births per woman) has declined steadily since 1977 from 3.3 births to 2.6 births, Utah's birthrate is still considerably higher than the 1.80 average of other states. This surge in birthrates is expected to continue, averaging 1.2 percent a year from 1987 to 1993, when the birthrate is expected to decline.

By the year 2010 Utah's school-age population is expected to increase to more than 65,000 children, a 15 percent increase from 1987. Given the fact that the 82,300 new jobs created in Utah between February 1983 and February 1987 did not meet the demand of the thousands of bright students graduating from the state's universities, state officials are concerned that Utah will not be able to provide enough jobs for future graduates. The governor estimates Utah needs 25,000 new jobs each year just to put the state's graduating classes to work and prevent this resource from becoming the state's most valuable export.

Utah high schools grant diplomas to almost 76 percent of the state's high school students. The national high school graduation average is 70.6 percent. Also, nearly 20 percent of Utah's high school graduates go on to finish college.

It is also true that most of Utah's revenues go to education. Utah ranks first in the

A wide spectrum of medical services, combined with a strong educational system, provides Salt Lake Valley with a healthy and well-educated life-style. Courtesy, Intermountain Health Care

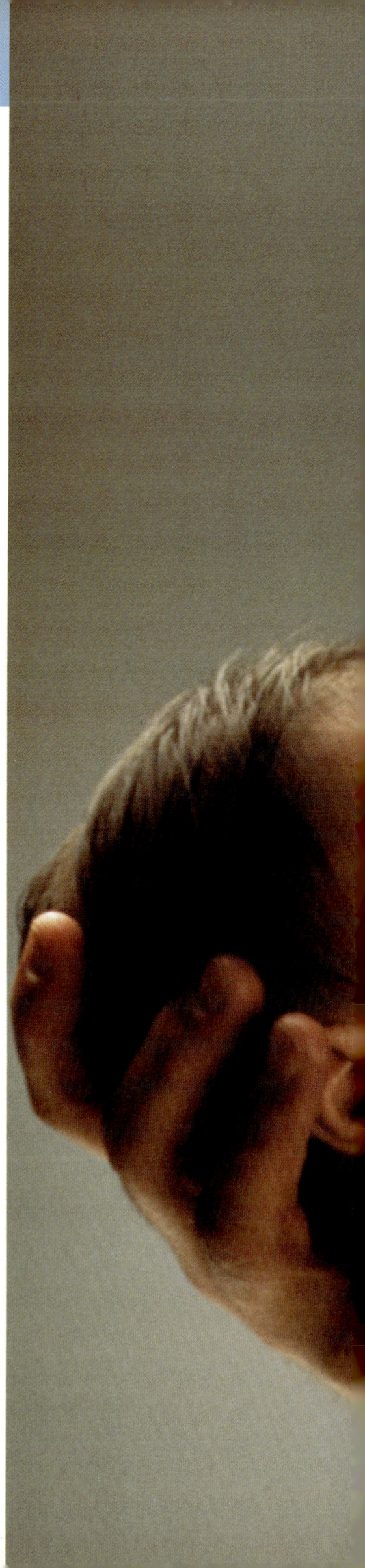

Right: Students of Lowell Elementary concentrate on a school project in the colorfully decorated classroom, as their teacher looks on. Photo by Stephen R. Smith

Below right: Hillcrest High School in Salt Lake City is one of the newer schools to be built in the Salt Lake Valley. Photo by Mark Gibson

Below: Education can often extend past the classroom and into the community, filling a vital need for residents. Volumes of information are available at the Geneological Library in Salt Lake City for those seeking to learn about their family trees. Photo by Mark Gibson

nation in taxes allocated for education. However, because of the state's unusually large student population, Utah ranks 48th in money spent to educate each child. In 1987 Utah's per pupil expenditure of $2,455 was 61.6 percent of the national average of $3,983 per pupil.

With an ever-increasing number of new students to teach (two students entering the system for each one who leaves), Salt Lake City's educators are constantly searching for more funds. To assist schools, the state passed a controversial tax increase in 1987, the largest in the state's history. Critics of the tax increase argued that businesses were not paying their fair share of taxes while reaping the benefits of an educated labor pool. "If businesses paid their fair share, it would raise our expectations of the education system, in turn increasing salaries and incentives for intellectuals to move here," says Mayor DePaulis.

However, supporters of the tax increase argued that this major effort by taxpayers will be rewarded in the future when graduates move into the work force. Furthermore,

Salt Lake City and all of Utah could only benefit from having a larger, better-educated work force than most of the rest of the nation. This indeed would be an enviable position if Utah's young graduates could find good jobs.

Despite alleged underfunding, Salt Lake City students outperform the nation on the American College Test (ACT) and rank first in the nation in advanced placement enrollment. Although Utah ACT scores dropped slightly in 1987, they still topped the national averages in all academic areas except mathematics. According to U.S. Secretary of Education William Bennett, "the state's ACT score has risen consistently since 1982, and overall, Utah has posted the nation's fourth largest ACT gain since 1982." Students in Granite School District in Salt Lake County, the state's largest district, scored even higher than other students in Utah and the United States in all subjects, including math.

Utah continues to lead the nation in the number of high school students who take advanced placement courses. Utah's program, which allows high school students to take courses for college credit, is a model for similar programs throughout the country. In 1987 Utah had 543 students per 100,000 population taking advanced placement examinations, compared to a national average of 162 students per 100,000 population. New York, which ranks second behind Utah for advanced placement examinations, had 293 students per 100,000 population taking the exams.

Utah has also initiated tougher high school graduation requirements and more rigorous college admission standards. "The data shows that when you increase expectations, you see a substantial difference in course-taking patterns," says David E. Nelson, director of evaluation and assessment at the State Office of Education. A recent study shows the percentage of Utah seniors taking foreign language, history, social studies, math, science, and art courses has steadily increased in the last four years. In 1984, for example, 11.6 percent of high school seniors were taking two years of a foreign language. In 1987 that figure jumped to 37.9 percent. Furthermore, results from a 1987 survey of high school seniors show that at least 80.4 percent of high school students plan to pursue some form of education after high school. Of all respondents, more than 46 per-

Brigham Young University in Provo is part of the fine selection of higher education institutions in the Salt Lake area. Photo by Stephen R. Smith

Both traditional and innovative methods of study are used in Salt Lake schools, promoting strong skills in all levels of instruction. Photo by Stephen R. Smith

cent planned to enroll in a four-year college, naming the University of Utah their first choice.

"Utah's public school system embodies much that makes Utah so unique—our tradition of large families, our historical commitment to the importance of education, our uncompromising support for rigorous academic standards, our expectations of high moral standards and strength of character for our youth, and our desire to make schooling available to all our citizens," states James R. Moss, state superintendent of public instruction, in his annual report. "Schools are an integral component of our society, a measuring rod of our social conscience and a basic source of our prosperity." An educated society benefits everyone by providing a solid foundation for higher-paying jobs, a quality life-style, and a prosperous community.

To maintain that standard of excellence, Salt Lake City schools use both traditional and innovative approaches. Utah's Core Curriculum standardizes the public school curriculum by defining teaching standards and outling course objectives for all grade levels. The Utah Core, established in the last three years, "calls for mastery of information, concepts, and skills in language arts, mathematics, science, social studies, the arts, informa-

Education sometimes leaves the classroom behind. The University of Utah's Division of Continuing Education offers hundreds of courses and workshops ranging from painting and psychology to the more adventurous, like this four-day canoe trip down the Green River. Photo by Stephen R. Smith

tion technology, healthy life-styles, and vocational preparation, with provision for electives and concentration in areas of a student's special interest." Utah leads the nation in course requirements in the fine arts and total credits for graduation.

Utah follows an outcome-based educational philosophy, which increases student learning by emphasizing and monitoring specific learning outcomes. "The emphasis is no longer on merely providing students with an opportunity to learn; the emphasis is now on the obligation to teach them," says Moss.

Students with special needs—those with learning disabilities or speech and hearing problems—are assured the same quality education as other students. More than 40,000 students with disabilities live in Utah and require these special education services. A team consisting of parents, teachers, administrators, and, when appropriate, the student develops an Individual Education Program (IEP) for the child. "These services must be delivered in the least restrictive environment appropriate to that student's needs," reads their annual report.

Public school officials and college educators regard extracurricular activities as an important component of a well-rounded education. Extracurricular activities teach cooperation, discipline, and sportsmanship, and therefore are encouraged. During the 1986-1987 academic year, nearly half of Utah's students participated in 20 different sports activities, forensic/debate, music, drill team, or drama competitions.

Learning in Salt Lake City is not confined to the classroom. Many of Utah's schools offer programs in the performing arts, which add another dimension to learning by stimulating a student's imagination and encouraging creativity and better communication skills. Each year Ballet West, Utah Opera Company, and the Utah Symphony perform at schools throughout the state. The state's Sterling Scholar program, now more than 20 years old, recognizes high academic achievements by high school seniors and provides scholarships to Utah colleges and universities. Several organizations, including KSL Television and the *Deseret News,* sponsor annual competitions in English/language arts, social studies, foreign language, visual arts, home economics, science, mathematics, music, vocational trades, and general scholarship categories.

Utah's Area Vocational Centers have also received national recognition. These vocational programs assist high school students and adults in career planning and vocational training. Students can take classes in computers, business, marketing, health, home economics, industrial arts, and agriculture. Courses are open-entry; for example, an unemployed plumber does not need to enroll for an entire semester program if only one class is needed.

Custom Fit Training works with business and industry and provides intensive training for company employees. In 1987 Custom Fit trained 1,871 individuals, 1,847 of whom were hired by Utah firms. The program provides funds and services for classroom courses or on-the-job training designed by the employer.

The Young Entrepreneur Program promotes the principles of free enterprise and is now a part of the secondary vocational program. Utah's high school principals select and recognize students who own and operate their own businesses while attending high school.

Utah Business Week, another program devoted to teaching the free enterprise system, is sponsored each year by the Salt Lake Area Chamber of Commerce. Begun in 1980, the six-day intensive workshop is designed to teach high school students and teachers in the state about free enterprise and business. More than 200 students representing 80 high schools are selected to attend each year. Most of the money used to sponsor the event is donated by Utah businesses. Dozens of volunteers from the business community serve as advisers, faculty, discussion leaders, and staff. Daily classroom sessions cover international economic systems, business strategies, marketing, advertising, deregulation, and government.

Students also spend a portion of the week actually running a mock business— profits, competition, advertising, and other areas—all simulated by a computer. "By the end of the week they truly appreciate the difference between earning a profit and going under," says Deborah S. Bayle, vice-president of administrative services for the Chamber of Commerce and coordinator of the Utah Business Week program.

Salt Lake encourages its young students to take advantage of the area's many secondary educational opportunities, expanding their range of knowledge and skills into the community environment. Photo by Stephen R. Smith

Other special programs offered through the public school system include law-related and citizenship education and alcohol, drug, and tobacco prevention. These courses are designed to prepare students to be responsible, law-abiding citizens and focus on values, ethics, patriotism, the Constitution, and the court system. Utah law also mandates that each public school student must take the alcohol, drug, and tobacco prevention program. This program helps students acquire skills that increase self-esteem, peer resistance, and communication. The course also teaches students how to make healthy and responsible choices about substance use.

The information age is not only changing the way people do business but is changing the way people learn as well. Utah schools are using several high-tech teaching tools, including instructional television and computers, to teach students throughout the state and across the country as well. Instructional television has two live broadcasts daily in advanced placement English and functional computing.

Computer-assisted instruction has been implemented in every school district in Utah and many programs are marketed to the entire country. Computer simulation and modeling help students visualize abstract concepts and analyze problems. With a computer, teachers can simulate the operation of a business or the problems of city government, for example.

Salt Lake City is also setting higher standards for the teaching profession. Utah's Career Ladder program, one of the best in the country, professionalizes the teaching field and improves teacher performance and retention. "It gives better teachers more responsibility and provides bonuses and incentives for quality work, much like you'd find in other professions," says Moss. "We started from the ground up in Utah, and other states have modeled their programs after ours."

However, studies show that throwing more money at the top of the system does nothing to improve what the student learns and retains in the classroom. "In the short term, no new tax increases for public education should be adopted," concluded a report by the Salt Lake Area Chamber of Commerce's Education Reform Subcommittee. The committee based this conclusion on the following:

•The rate of increase in net new enrollment is decreasing. If projections are correct and current trends continue, the rate of increase of net new enrollment should ap-

80

This high school student works diligently on her class assigment.
Photo by Stephen R. Smith

proach zero by the early to mid-1990s.

•Utah citizens are taxed at high rates compared to other states.

•In 1987 the Utah legislature enacted the largest tax increase in Utah's history, yet the state budget did not increase. Most of the revenues raised replaced revenue shortfalls.

•Utah's funding of the public schools has grown continually since 1976-1977.

Utah's business leaders are not dismissing the importance of education just because they oppose further tax increases for an already taxed-out population. Further taxation for education is a touchy issue. People who oppose further tax increases are not anti-education. Rather, many believe Salt Lake should give its young people the best education the state can afford.

The Chamber's recommendations for solving the education crisis include maximizing classroom capacity in existing schools instead of building new ones, conducting extracurricular activities after school hours with locally raised revenues, and consolidating school districts.

Salt Lake County is divided into four school districts: Murray City (6,079 students), Jordan (59,107 students), Salt Lake City (24,503 students), and Granite, Utah's largest school district and the nation's 37th largest (70,237 students). Since 1959 the Salt Lake City School District's enrollment has steadily declined as more people have moved to the southern part of the valley, resulting in 25 school closures. The most recent casualty came last year when the district closed the 57-year-old South High School. Boundaries were changed to balance enrollments among the district's three remaining high schools according to size, median income, minority enrollment, and achievement level.

More and more families with young children are moving to the southern and western areas of the valley: Sandy City, West Jordan, South Jordan, West Valley, and the Taylorsville/Bennion area. In some of these communities, the median age is 19 and the average student-teacher ratio is 26 to 1.

During the 1970s student enrollment at the Jordan School District roughly doubled, increasing at an extraordinary rate of more than 7 percent a year. Since 1980, enrollments in the Jordan district have increased 3.77 percent a year.

In an attempt to trim the school budget, the Utah legislature approved two

cost-cutting measures: First, by the 1990-1991 school year, all elementary schools in Utah must implement extended-day or year-round schooling. Second, all school facilities must be running at 70 percent capacity. "When we're forced to set priorities, creative things can happen due to the economics," says Moss. "The state will shut down schools not 70 percent filled, unless it is in a remote or decreasing-population area."

Most Utahns prefer additional methods of solving the financial troubles of Utah's schools. A recent poll found that Utahns rejected a four-day school week but believed a "head" tax charged to people with children who use the system would be fair.

Salt Lake City's many private schools do offer an alternative to public education. Parents may choose from a variety of specialty and parochial schools, each with its own philosophy and liberal arts curriculum.

Enrollment in private and parochial schools has increased steadily over the last 10 years, but most of the growth has been seen in the last three years, a trend that parallels increasing concerns about classroom sizes in Salt Lake's public schools. Fourteen private elementary and secondary schools are located within the valley. Private school enrollments account for only 1.3 percent of Utah's total school enrollment, which in 1986-1987 was 415,994. Private school enrollment, by comparison, was 5,866.

Salt Lake's largest private school, Judge Memorial Catholic High School, was originally built as a miner's hospital in 1921. Today, Judge is a four-year coeducational high school, the only Catholic high school in the Salt Lake Valley. The school's curriculum offers 175 courses and a wide range of athletic activities. Operated by the Roman Catholic Diocese of Salt Lake City, Judge is committed to teaching Christian values. The school's student body comes from all faiths (25 percent are non-Catholic) and areas of the valley.

Judge's student-teacher ratio is 18 to 1. The school's academic accomplishments are impressive. In the last five years Judge graduates have ranked in the 93rd percentile composite in the ACT testing program. Of the 53 seniors who took advanced placement exams last year, 90.6 percent earned passing scores compared to the national average of 68.4 percent. Ninety-two percent of Judge graduates go on to college.

The Catholic elementary schools include kindergarten to eighth grade and feed into Judge and St. Joseph High School in Ogden. Tuition in the elementary schools ranges from $575 a year (for Catholic students who are subsidized by their parishes) to $850.

Rowland Hall-St. Mark's School was founded by the Episcopal Church, but now has adopted a pluralistic approach to religious instruction. Even though the school opened a second campus four years ago, it still has a waiting list. Enrollment at the school's two campuses is around 800, up from 331 in 1982. The lower school includes kindergarten through sixth grade, in addition to a preschool; the upper school includes grade seven through high school. Rowland Hall-St. Mark's school has an impressive academic record. Almost all of its high school graduates go on to college, and, at times, as many as 30 percent of the senior class have been National Merit semifinalists. The Rowmark Ski Academy, founded in 1981, has become internationally known. Over 50 percent of Rowmark team members rank among the best 50 skiers in the nation in their respective age groups.

Four years ago, the Lutheran community organized Salt Lake Lutheran High School as an extension of its elementary system. The valley's 10 Lutheran congregations fund 10 percent of the school's budget, another 60 percent comes from tuition, and fundraising activities make up the remaining 30 percent. "About 50 percent of our 60 students are Lutheran. The others come from all faiths and all parts of the valley, even Park City or Provo," says principal Eugene Kolander. "We have a real mix of denominations. You can bet our religion classes are sure interesting." Right now, classes meet in a former church facility that accommodates 80 students. School officials expect to build a new facility, perhaps in the southern part of the valley where the population is growing rapidly. Kolander believes Salt Lake is an ideal city for private schools. "We'll see a dramatic increase in the number of private schools if current population trends continue," he says. "When you consider the tax relief when a child transfers from public to private,

Facing page, top: Besides teaching young children the necessary academic lessons, Salt Lake's educational system provides each student with the opportunity to learn about teamwork and friendship with their schoolmates. Photo by Stephen R. Smith

Facing page, bottom: Classmates at Judge Memorial Catholic High School, Salt Lake's largest private school, confer about their studies. Photo by Stephen R. Smith

a private education is actually less expensive."

Recently opened in Sandy, the Waterford School offers a liberal arts curriculum combined with a computer literacy program. School officials believe that today, technology must be a part of a child's education; therefore, the school offers many hands-on computer classes. The school includes nursery school through fifth grade. A new middle school will graduate its first seniors in three years.

Salt Lake also has several schools for gifted and talented children. Realms of Inquiry's admission standards, however, are fairly open and invite any child who is eager to learn. Tuition ranges from $2,300 for kindergarteners to $3,500 for high school students.

HIGHER EDUCATION

The greater Salt Lake area is a center of higher education, with four major universities and the world's largest private college, Brigham Young University, located along the metropolitan corridor.

The oldest state university west of the Missouri River, the University of Utah, originally known as the University of Deseret, was established by pioneers in 1850 just three years after they entered the valley. The "U," as it is affectionately referred to, today has an international reputation for academic excellence and attracts students from all 50 states and 65 foreign countries. It is a sophisticated, urban institution—really a city within a city, housing three residence halls, restaurants, cultural halls, elegant theaters, 27 tennis courts, 15 racquetball courts, four squash courts, an indoor running track, four gymnasiums, three swimming pools in one natatorium, a weight room, a nine-hole golf course, three outdoor playing fields, movie theaters, and even a 10-lane bowling alley. In addition, student services include a career research library, counseling, tutoring, a credit union, student health facilities and insurance, computer facilities and software, honors programs, and foreign exchange and international experiences.

The wooded campus covers 1,520 acres on the sloping foothills overlooking downtown Salt Lake City. More than 25,000 students attend classes here, graduating in one of 64 majors. The university has the fourth largest enrollment among 24 colleges and universities throughout the Mountain West. Nearly 85 percent of the university's students work at least part-time. "This campus is one of the prettiest I've seen. It looks and feels like a college, not an institution. Here you see students with backpacks among a mixture of modern buildings and ivy-covered, turn-of-the-century buildings. It just feels right," comments a business student from California. "It's a beautiful place. If you like

skiing and nature, there's no place like it," says a mechanical engineering student from Brooklyn, New York.

The University of Utah is one of the top institutions in the eight-state Mountain West and the leading university in the Intermountain region of Nevada, Utah, Wyoming, Idaho, and Montana. Of all universities in the Intermountain West, it has the most revenues and the largest research program. It ranked third in eight states in total revenues: $311.5 million in 1986. It also ranked third in total non-tax revenues generated: $203 million. The University of Utah is also Salt Lake's largest employer with approximately 16,000 employees. Combined, the University of Utah and Research Park paid $321.6 million in wages in 1987, second in Utah only to Hill Air Force Base in Weber County. In 1986-1987, taxes paid toward university funding amounted to $66 per resident. The university, in turn, generated more than $200 that year for every man, woman, and child. Considering all expenses, including goods and services purchased with university revenues, the institution contributed one billion dollars to Utah's economy.

The colleges within the university and the Division of Continuing Education offer a wide variety of graduate, undergraduate, and academic programs on a quarter system. The Division of Continuing Education offers more than 950 high school and college-level courses in the evenings, on weekends, and during the day, ranging from river rafting to writers' workshops. Estimated university resident tuition for three quarters, 12 credit hours per quarter, is $1,221 plus $510 for books and supplies. The estimated tuition for a nonresident is about $3,485.

The university has traditionally been supportive of its faculty. Said one university professor, "Together we open minds, seek answers and go wherever questions lead us. Teaching is essentially learning with your students, stimulating them, offering resources and encouraging them to follow their intellectual honesty."

The university also has one of the finest dance programs in the country. Students are trained in performance and choreography. Many students go on to teach or direct

Above: Marriott Library at the University of Utah holds nearly 2 million volumes on its shelves. Photo by Stephen R. Smith

Top: The University of Utah enrolls more than 25,000 students at its campus near the foothills, above downtown Salt Lake City. Photo by Stephen R. Smith

dance classes or join professional dance companies. The university dance program has nurtured the careers of many dancers who are now part of professional companies such as Ballet West, Repertory Dance Theatre, and Ririe-Woodbury Dance Company.

According to an Association for Communication Administration survey of top communication scholars around the nation in 1983, the University of Utah's doctoral program in communications ranked 10th among 60 programs in the nation in overall quality, 8th in mass communication, and 5th in organizational communications.

The university's business, engineering, law, and medical schools have all received wide recognition.

The university also has some of the Intermountain area's foremost libraries and research centers (Marriott Library has nearly 2 million volumes), a public television and radio station, and two museums. The Utah Museum of Fine Arts displays year-round exhibits, traveling art collections, lectures, and films. The Utah Museum of Natural History showcases more than 175 exhibits on geology, biology, ecology, and man.

Students aren't the only fans of the U's sports teams. In this respect, Salt Lake City is a "college town." Just ask any businessman if you want to know who won the game last night, but be prepared for a quarter-by-quarter recap. Rivalries between the University Utes and the Brigham Young Cougars—and their respective followers—are legendary among longtime residents. The women's teams deserve some recognition, too. The Lady Utes gymnastics team was NCAA champion for six years.

The University of Utah campus is lined with maples, sycamores, fruitless mulberries, and even the unusual Chinese poplar or Japanese zelkova. The entire campus, with 8,000 trees, is the State of Utah Arboretum. In May the white lace of the ornamental pear and bright pink blossoms of the flowering peach decorate the grounds around the old stone buildings along President's Circle, transforming this section of Salt Lake into a postcard scene from New England. The arboretum is part of the Immigration Visitors District which links the museums and nearby Hogle Zoo, Pioneer Trails State Park, "This Is the Place" monument, and Fort Douglas in a coordinated marketing effort.

Not far from the university, in a neighborhood of ivy-covered brick tudor homes along streets with towering maples, is Westminster College. A private college, it is the only small liberal arts college in Utah. The college was founded in 1875 as a Presbyterian preparatory school. Westminster offers 25 bachelor's degrees in science, education, business, nursing, and the arts, as well as two master's programs. Administrators expect that by 1990, between 1,800 and 2,100 students will be enrolled. The student-faculty ratio is 17 to 1. Almost 75 percent of Westminster's students are adults returning to finish a degree. The school operates on a semester system, and a full-time program of 12 credit hours costs $5,200 a year.

Westminster is situated on 2,300 acres. The Gothic buildings are obscured by columns of maples, cottonwoods, and box elder. Emigration Creek runs through campus, where the college operates a bird sanctuary.

Salt Lake Community College, formerly known as Utah Technical College at Salt Lake, has more than 10,000 students enrolled in 60 programs on three campuses. In addition to vocational and occupational trade curricula, Salt Lake Community College offers associate of arts degrees and courses in early childhood development, pre-engineering, automated systems technology, manufacturing technology, computer science, criminal justice, physical education, and foreign languages, in addition to aerospace studies, architectural drafting, electronics, and many other subjects. The average age of students at the college is 29, and the student-faculty ratio is 23 to 1.

High school graduates and older Utahns may also choose from several business schools such as Bryman, LDS Business College, or the University of Phoenix.

Former mayor Ted Wilson says:

Every city has its own set of ailments. We need to let people know why we live here. Our heritage is dedicated to education. We have a wealth of engineers, electricians, medical scientists. These people are our most important product; that's what we'll sell in the future in our role in the world market. With the rest of the nation entering a labor short-

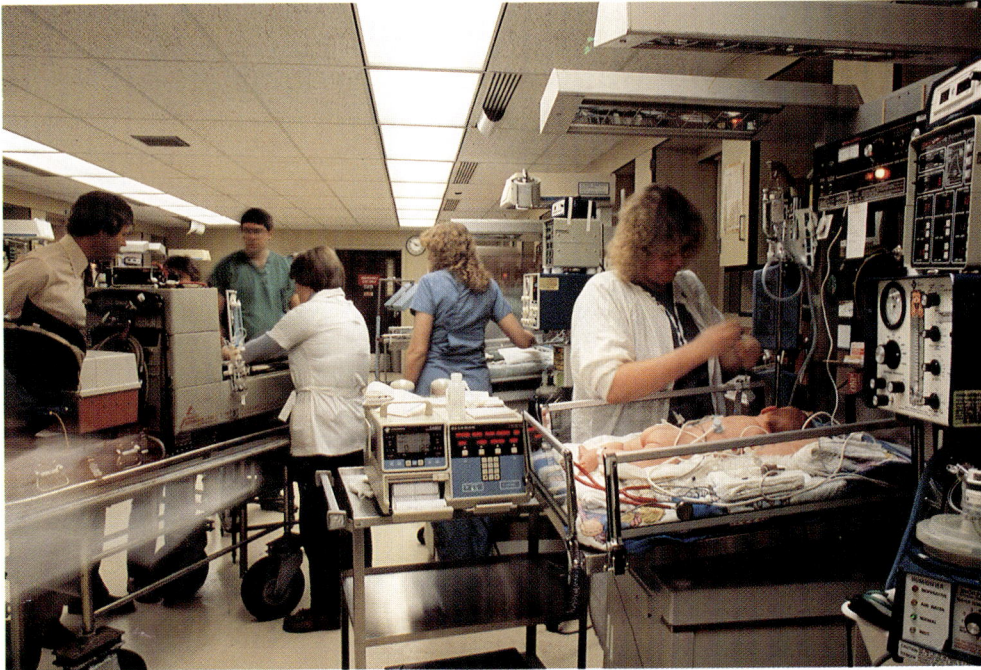

The quick and efficient staff members of the newborn intensive care unit at Primary Children's Medical Center in Salt Lake City respond swiftly to the needs of their young patients. Courtesy, Intermountain Health Care

age, there's a lot of expansion power there. We're unique in terms of our human resource development. Salt Lake is small enough, we can improve the education system from the parents up. From that base, the economics will take care of themselves.

A HEALTHY PLACE

One of Salt Lake City's fastest growing and most stable industries is health care. Health care service in Utah is a $2-billion-a-year industry and is growing about eight to nine percent each year. In 1985 approximately 51,000 Utahns earned $891.8 million working in some type of health-related business. Since 1983 the number of physicians and other medical personnel has increased over 4 percent each year. Diverse and specialized health care programs offer a wide range of treatment from routine checkups to emergencies. Skilled professionals operate trauma and burn centers, CAT scanners, and emergency transport planes. Paramedics respond to 911 emergency calls within three minutes in downtown Salt Lake City. Dental implant procedures have replaced the need for dentures. Better vision is possible thanks to laser surgery. Volunteer donors maintain constant supplies of blood at blood banks. Sports medicine and rehabilitative therapy have increased as amateur and professional sports thrive in Salt Lake.

The East Bench Corridor, as some call it, is the very heart of the valley's health care system. This area includes the University Health Sciences Center—commonly called the Medical Center—as well as LDS Hospital, Shriners Hospital, Research Park, the Western Institute of Neuropsychiatry, Veterans Administration Hospital, and Holy Cross Hospital. All are located either in or near downtown Salt Lake City. Primary Children's Hospital will soon move from its 36-year-old facility in the Avenues to new quarters adjacent to the University Medical Center. Better access to university resources and health professionals will enhance the hospital's role as one of the world's foremost pediatric care centers.

When the George S. Eccles and Delores Dore Eccles Institute of Human Genetics is completed in July 1990, this quadrant of the valley will rank among the nation's top medical research and health care centers. Researchers here hope to unlock the mysteries of genetic disorders that cause disease—such as diabetes, cancer, and cardiovascular disease—and find a treatment.

The Howard Hughes Medical Center has contributed $11 million toward construction of the eight-story genetics research and basic science building, now under construc-

Faster and more efficient methods of communication between physicians have helped to improve health care in Salt Lake. Courtesy, Intermountain Health Care

tion and estimated to cost $24.5 million. The Eccles Foundation, organized in 1969, has pledged one million dollars a year to assist the Hughes Institute in building the facility, according to Spencer F. Eccles. The foundation has also committed to an additional one million dollars a year for five years, beginning in 1989. This money will be used to support visiting scholars, fellowships, postdoctoral candidates, and other programs. The late George Eccles was a member of the Eccles Foundation board and chairman of First Security Corporation.

The Hughes Medical Institute has two laboratories at the University of Utah, one in the Department of Biology and a genetics laboratory at the School of Medicine. These two laboratories are among 14 other laboratories in U.S. medical schools (including Harvard, Yale, Johns Hopkins, and Columbia universities) that form the Hughes Institute.

According to a University of Utah newspaper, officials at the Hughes Institute are excited about the new Utah facility. "We regard our investment at the University of Utah as the kingpin of the key initiative to spur the mapping of human genome [the identification of the specific genetic defects in human disease]," says Donald S. Fredrickson, institute president and chief executive officer. "No other location in the world could provide the data base from which scientists skilled in molecular genetics techniques can develop this information which is of great importance to human beings everywhere."

Internationally respected Dr. Raymond White, a medical geneticist, will supervise research at the center. "Utah is unique as a 'world resource' for the study of genetics," says White in the article. He has recently discovered a new genetic marker near the gene that causes cystic fibrosis, the most common and fatal of all inherited diseases in the United States. "Utah's vast genetic records, large families and stable population that has strongly supported medical research are resources that exist nowhere else," says White in the article.

"With the new Eccles Institute of Human Genetics, the University of Utah will be one of less than a half-dozen centers for exploring human genetic diseases," says Dr. James Brophy, the university's vice president of research. "We'll be at the forefront of the next breakthrough in the health sciences."

"With the new Primary Children's Hospital, the expansions at VA Hospital, Holy Cross, and LDS hospitals, and the new Genetics Research Center, Salt Lake will be the true Rochester of the West," says Fred Ball, president of the Salt Lake Area Chamber of Commerce.

The University of Utah is the only teaching hospital among the major universities in the Intermountain West. University Hospital is the major medical, referral, research, and training center for all five states except a portion of Nevada.

The University of Utah Health Science Center employs more than 3,600 people and has an annual payroll of more than $90 million. Established in 1975, it includes the colleges of health, nursing, and pharmacy; School of Medicine; University of Utah Hospital; Spencer F. Eccles Health Sciences Library; Student Health Service; and Regional Dental Education Program.

Scientists at the center are trying to unlock the secrets of molecular genetics, population genetics, and the genetics of clinical disease. Within the next decades, research teams hope to piece together the genetic puzzle of tumors, especially those afflicting young children. Cardiologists are studying the causes of hypertension, while other scientists are looking for answers regarding hemophilia.

In addition, the Department of Psychology has formed a teaching and research network with campus medical researchers. Together, they study the inter-relationship between the psychological and biological characteristics that shape human beings. This network includes experts at the Veterans Administration and LDS hospitals. The team also has studied amnesic diseases, such as Alzheimer's disease.

The university's Department of Health programs focus on physical therapy and physical education. Specialists treat victims of muscular dystrophy and study the disease's patterns and progress. Doctors are optimistic that a drug will soon be discovered that will treat or even arrest muscle deterioration in muscular dystrophy victims, who are mostly children. University scientists have developed a method of charting the disease's progress. Funded by the Muscular Dystrophy Association, these studies show that muscular dystrophy leads eventually to total muscle deterioration. Other muscle disease clinics around the country have implemented the University of Utah model.

Often working together, these health care centers develop innovative health care programs for the community. For example, the university and the Veterans Administration hospitals jointly study the benefits of exercising, expecially its effects on the elderly.

One of the nation's first teaching nursing homes was established in Utah in 1982. It is a model facility that provides clinical care, education, and research. As the majority

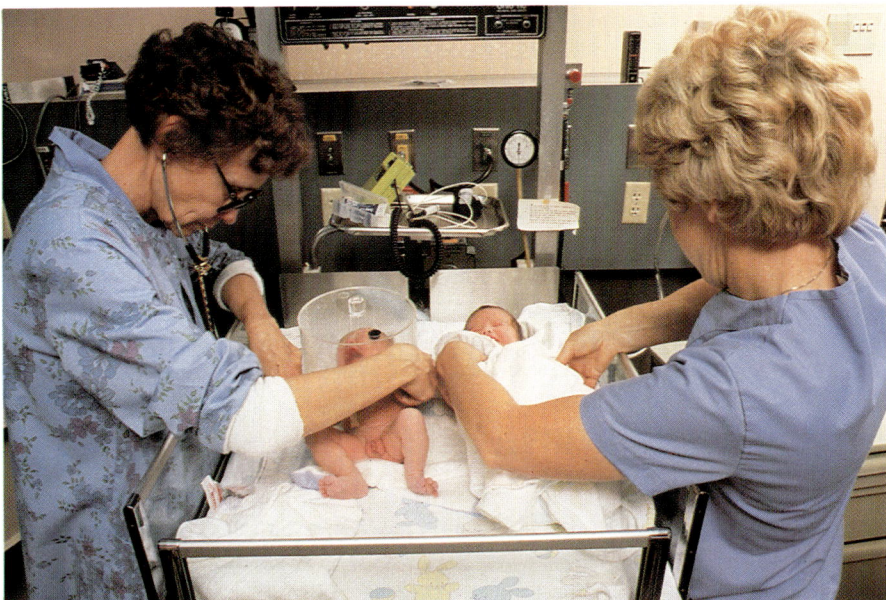

Obstetric nurses tend to newborn twins, just delivered by emergency cesarean section, at the Cottonwood Hospital Medical Center in Murray. Courtesy, Intermountain Health Care

of America's population grows older, more nursing homes will be needed to care for elderly men and women who are living longer than ever before. "Linking nursing homes with colleges that prepare geriatric nurses enhances the quality of care for patients," says Dr. Margaret Dimond, associate professor of nursing. The Department of Health and Human Services Administration on Aging has funded the Intermountain West Long Term Care Gerontology Center at the College of Nursing. Several university departments, state agencies, and volunteers are working to improve long-term health care in Utah, North and South Dakota, Colorado, Wyoming, and Montana.

Academic health care centers, such as the University of Utah Health Sciences Center, make enormous contributions to improving and maintaining the quality of health care in a community. Teaching hospitals generally attract some of the best physicians in the country.

Other area hospitals in the central and southern sections of the valley include Alta View in Sandy, St. Mark's in Holladay, Cottonwood in Murray, Pioneer Valley in West Valley City, and Holy Cross Jordan Valley in South Jordan. Most hospitals are owned and operated by Intermountain Health Care, Inc. (IHC). IHC acquired 15 facilities in various states of disrepair when the Church of Jesus Christ of Latter-day Saints sold its hospitals in the Mountain West. IHC immediately set out to modernize the facilities. In 11 years IHC has spent $430 million to replace buildings and purchase equipment. It has also developed a core of central services, such as billing, data processing, purchasing, and insurance, to streamline their operation, thereby cutting costs. IHC is a full-service health care provider with 23 hospitals and 260 affiliated hospitals, mostly in Utah and the Intermountain West.

IHC subsidiaries provide additional support services, including home health care, women's health care, ambulatory surgery, psychiatric-behavorial health, occupational health, rehabilitation therapy, neighborhood clinics, and nonprofit care for indigent patients.

Regional hospitals frequently cooperate to provide special, innovative programs to the community. LDS Hospital and the University of Utah Medical Center, for example, form the Intermountain Trauma Complex. LDS Hospital's Life Flight program rescues and transports critically ill or injured patients to trauma units and serves residents within a 100-mile radius of Salt Lake City.

Utah was one of the first states in the nation to permit alternative health care systems to compete in the open marketplace. Currently the valley has 20 alternative health care systems which serve more than 300,000 residents. These health maintenance organizations estimate that the number of facilities will double in the next few years. Employers and families will be able to make many health care choices and save money.

For people who need treatment but who do not require hospitalization, alternate health care facilites and emergency centers are located throughout the valley. There are several licensed home health agencies offering specialized care to the elderly and mentally handicapped. The Salt Lake County Mental Health office operates five outpatient units, all of which provide diagnostic evaluation for alcohol and drug abuse, and couples and group therapy. In one year alone more than nine new psychiatric care centers were established along the Wasatch Front to care for individuals suffering from depression, adolescent problems, or drug or alcohol abuse.

Salt Lake hospitals have earned a reputation for quality care and efficiency. For example, its hospital admissions per bed rate (a measure of efficiency) is 44.1 compared with the national rate of only 34.6. Costs in the mid-1980s were 14.7 percent lower than the national average.

The high caliber of medical research being conducted in Salt Lake's hospitals and universities has made the area a worldwide leader in quality health care, disease research, and health education. Like the rest of the country, Salt Lakers are hooked on health. More than 35 active health-related associations and foundations participate in the community. Apart from fun runs, marathons, and wellness workshops, educational programs are also offered in the workplace. "Healthy Utah," a division of the State Department of Health, periodically conducts seminars for employers to help improve employee productivity. In addition, public and private agencies conduct a "Smoking on the Worksite" conference, and promote the "Baby Your Baby" prenatal awareness campaign.

The breakthroughs in medicine occurring in Bionic Valley are one of the reasons for Salt Lake's healthy population. These medical advances have resulted in higher life expectancies, lower mortality rates, and less sickness and disease in the community. Utahns enjoy excellent health, shorter hospital stays (5.4 days compared to the nation's average of 7.7 days), and longer lives than other Americans. Because Salt Lake residents live a healthy life-style, they have lower-than-average rates for the top four leading killers: heart disease, cancer, strokes, and accidents. A high percentage of Utahns are nonsmokers, and the average life span is 72.90 years, the nation's third highest longevity rate. Utah's residents, for the most part, advocate clean and wholesome living.

A healthy life-style is important to Salt Lake Valley residents. Photo by Stephen R. Smith

CHAPTER SEVEN

THE GOOD LIFE

Recently, on his first visit to the Beehive State, Mayor Ed Koch of New York remarked, "Salt Lake City is the cleanest city this side of heaven."

It often takes such a reminder from visiting dignitaries, friends, or business associates for Salt Lakers to appreciate the qualities of their community: the clean streets and parks, buildings free of handbills and graffiti, an excellent bus system, quiet residential neighborhoods, and an endless variety of things to see and do in and around the valley and nearby mountains. "Salt Lake City residents enjoy the good life in a city that otherwise has a remarkable absence of those big-city problems which have come to plague other fast-growing regions," observed a travel editor in the *San Jose Mercury News*.

SOMETHING OLD, SOMETHING NEW

Strolling along South Temple, Koch admired the architecture along this historic street. Some of the city's grandest and oldest architecture adorns South Temple: Temple Square, the Hotel Utah, the Lion House (once the home of Brigham Young), the Eagle Gate Arch over State Street, the Elks Building, Cathedral of the Madeleine, First Presbyterian Church, and the Kearns Mansion all pay tribute to the city's glorious past.

Although many historic buildings in this section of the city have been preserved, others, such as the old Utah Bank building, have not. Today many business leaders recognize the cultural benefits of preserving what is uniquely Salt Lake. "Salt Lake has to have something besides great snow; we need to preserve our history," says Merlene Leaming. "Once it's gone, it can never be re-created. Why do we travel so far to see Williamsburg, Virginia; Boston; Washington; or Europe? We go to learn of our past and capture a bit of history."

Temple Square, Salt Lake's most famous tourist attraction, is also part of America's heritage. Nearly 700,000 people visited the grounds in July 1988. It is the city's most notable landmark and perhaps the best evidence of Salt Lake City's devotion to its heritage. Temple Square is a tribute to those early pioneers who, immediately after entering the valley, selected the location for their temple. Construction began on February 14, 1853. Cut granite slabs taken from the mouth of Little Cottonwood Canyon, some 20 miles to the southeast, were hauled back to the site on wagons. These granite slabs form the walls of the giant structure. Forty years later, on April 6, 1893, they dedicated the Salt Lake Temple.

Up the street from Temple Square, Father Lawrence Scanlan—the first bishop of the new Roman Catholic Diocese of Salt Lake—acquired land for a new cathedral to serve the growing numbers of Roman Catholics settling the area. On July 4, 1899, groundbreaking ceremonies were held on the steep hillside. But construction was slow since Father Scanlan preferred the "pay as you go" method. The new Cathedral of St. Mary Magdalene was finally completed and dedicated in 1909. It was rechristened Cathedral of the Madeleine in 1918. The cathedral is a magnificent example of Gothic and Romanesque architecture. The building sits on a hillside at the base of the city's historic Avenues residential area. The arched doors of the entrance lead into a sanctuary with a high vaulted ceiling and ornate altars. The altars are made of Utah marble. Scanlan spent $300,000 to construct the cathedral and $43,800 to furnish it. Nearly $88,000 was collected from Catholics in Utah and Nevada.

In 1917 work began on renovating the interior of the enormous cathedral. John Comes, "one of the most gifted architectural designers America has ever known," directed the project and "conceived an interior of rich color and magnificent design." America's leading artists carved

Ballet West's stunning production of Anna Karenina *features the dancing talents of Bruce Caldwell and Lisa LaManna. Photo by Jack Mitchell/Ballet West*

Historic South Temple Street features many excellent examples of significant Salt Lake architecture. Photo by Stephen R. Smith

Below: The colorful gardens and stunning sculpture at Temple Square are proud reminders of Salt Lake's continuing devotion to its heritage. Photo by Audrey Gibson

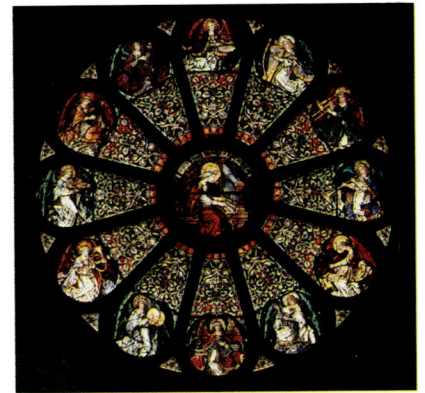

Above: The large rose window in the Cathedral of the Madeleine was designed in Bavaria, West Germany, in the tradition of the rose window of the Cathedral of Notre Dame in Paris. It portrays St. Cecilia, patroness of music, surrounded by 12 angels and ancient musical instruments. Photo by Stephen R. Smith

Left: The ornate interior of murals, columns, stained glass art, and statues makes the Cathedral of the Madeleine one of the most grand places of worship in the country. Photo by Stephen R. Smith

Above: A traditional market setting entices shoppers to browse through Trolley Square. Photo by Mark Gibson

statues, painted murals, and designed stained glass windows depicting the story of Christ. Today parishioners and visitors still admire the artisans' work.

Another fine example of historical preservation in Salt Lake City is the Denver/Rio Grande Railroad Depot, headquarters of the Utah Historical Society. When the railroad abandoned the building, they sold it to the state of Utah.

Just south of the central business district is Trolley Square, the second most visited attraction in the city. The old trolley barns now house quaint shops such as the Basket Loft, Feathered Friends, and an antique shop; strolling along cobblestone streets visitors can browse through apparel boutiques, have a pint at The Pub, or dine at any number of sidewalk cafes.

As Utah's capital city, Salt Lake City must now face the challenge of preserving the city's past while making way for the future. Recently a team of architects, the Regional/ Urban Design Assistance Team (R/UDAT), visited Salt Lake to evaluate the city's layout and to recommend a plan that would integrate older, historic buildings with modern structures. The team offered several suggestions, including defining Salt Lake City's boundaries or gateways, giving the city a "sense of place, and defining who we are." According to the plan, current sections of the city would be expanded as the arts centers, government centers, shopping/retail corridors, or commercial sections. The team also noted that downtown Salt Lake City has too many concrete "car jails," which are eyesores and encourage further traffic problems. The team also recommended preserving older buildings along Main and State streets. They discovered, though, that residents seem ambivalent about older buildings that give "fabric and texture to the downtown."

Carrying out this plan means preserving older buildings in the downtown area and restraining new construction east of the central business district. The team's objective is to preserve the city's greatest asset—its surrounding mountain vistas. These are what set this community apart from the dismal sameness of other cities. The city's beauty should not be obstructed, warned the R/UDAT. Good planning also involves "just saying no" to some projects, the team stated. Their idea is to design a city for its residents, plant-

ing more trees, creating street medians for safe pedestrian crossing, and encouraging more shop and sidewalk cafe business along the boulevards.

A city's character is reflected in its architecture. According to historians a building does not necessarily have to be a major structure to be architecturally or historically significant. A house representing an architectural style or period or a building designed by a particular architect has historical value. "People should ask, 'does it tell us something significant about our past?'" says Roger Roper of the Utah State Historical Society. Residents and developers need to realize that there is room for the new but not at the expense of the old. "Who knows, perhaps 70 years from now historians will look back at our modern architecture and find something of significance."

A TRIBUTE TO THE ARTS

Culture. It is the one immeasurable gift the city can offer people and businesses that the country or suburbs cannot. To truly appreciate living in Salt Lake City, residents should attend a performance of *Giselle* by Ballet West, take in a play at Pioneer Theatre, or enjoy an evening concert given by the Utah Symphony. "The arts not only feed the soul, they spark the imagination. The measure of a civilization is the creativity of its people. Through the arts, Utah can inspire its men, women, and children to reach for the very best," writes Governor Norman Bangerter.

Utahns' appreciation of the visual and performing arts dates back to the days of the pioneers. When the Mormons fled their homes in Illinois, they left behind all but the bare essentials needed to begin a new life in the West. Yet many of these pioneers hauled pianos and other musical instruments through the plains and over the Rocky Mountains to their new home. Brigham Young's dream was to create a civilization—not just a settlement—somewhere in the isolated Great Basin. In that quest he searched for artisans, teachers, historians, writers, and musicians to join him in his kingdom. He sent five Mormon artisans to Paris to study under the great impressionists; upon their return, they painted murals for the Temple. The world-famous Mormon Tabernacle Choir has its pioneer roots in traditional choral singing as a volunteer ensemble.

In 1899 the third Utah legislature created the country's first state arts agency to encourage the growth of the arts. Today the Utah Arts Council carries the arts into every area of the state. It remains a vanguard of the arts in the West. Utah is ranked as one of the top states in the country for per capita funding of the arts.

The Utah Endowment for the Humanities issues grants for history lectures, reading programs, and other humanities events throughout the state, bringing the works of scholars, writers, historians, and philosophers to the public. The organization recently sponsored a public television program entitled "Life Issues" that discussed capital punishment and abortion and was moderated by Pulitzer prize-winning playwright Arthur Miller. "We're here to elevate the quality of life in our state," says Delmont Oswald, director of the Utah Endowment for the Humanities. "Basically, everybody lives, eats, and breathes. We want to contribute to our existence and make our lives more enjoyable through education. Learning is fun, and doesn't stop with a formal education; it's a lifelong journey." For the past six years the Utah Endowment for the Humanities has been recognized as one of the top eight humanities programs in the country.

Utah can indeed boast of a wealth of arts and humanities programs. Salt Lake City has a world-class orchestra in the Utah Symphony; an internationally acclaimed ballet

The world-famous Mormon Tabernacle Choir is a mainstay of Salt Lake's cultural heritage. Courtesy, Utah Travel Council

company in Ballet West; two nationally known modern dance companies in Repertory Dance Theater and Ririe Woodbury, each about 25 years old; and a grand Utah Opera Company.

Salt Lake's Ballet West, now 26 years old, is the only professional ballet company between St. Louis and San Francisco. "We have the biggest ballet company—over 40 dancers performing over 44 weeks a year—in the least populated of ballet cities," says Trevor Cushman of Ballet West. "Other cities like Dallas, Seattle, Cincinnati, or Philadelphia have smaller dance companies with a population base two to three times larger than Salt Lake's." Ballet West has grown from a University of Utah dance program and civic dance group to a nationally recognized professional dance company that has performed twice at the Kennedy Center. The troupe's most famous production is Tchaikovsky's popular *The Nutcracker*. Other presentations include *Romeo and Juliet*, *Giselle*, and *Anna Karenina*.

At home in Salt Lake, Ballet West rehearses and stages its productions in the historic, renovated Capitol Theatre, accompanied by one of the nation's top orchestras, the 49-year-old Utah Symphony.

The Capitol Theatre, built in 1913, has outlived numerous natural and economic trage-

dies and has seen the evolution of the dramatic arts played within its walls. In the late 1920s the Capitol Theatre pre-sented silent movies accompanied by a Wurlitzer pipe organ. The advent of sound equipment in 1929 introduced the first talkies to Salt Lake City. In 1947 remodeling plans called for up-dated equipment, larger screens, and new seating. Sadly, on July 4, 1949, a fire raged through the building, killing one usher.

The 1950s and 1960s brought big-name Broadway musicals to Salt Lake City. *The Sound of Music, Carnival,* and *My Fair Lady* played to sellout crowds at the Capitol Theatre and rivaled the ac-claim given to such modern classics as *CATS* or *A Chorus Line.*

However, as small, neighborhood movie theaters began to spring up throughout the valley in the early 1970s, the Capitol Theatre was unable to compete and closed its doors after its final performance of *Shenandoah.*

"It's still hard to believe that this historic theatre was two weeks away from becom-ing a parking lot," exclaims Doug Morgan, the theater's stage manager who began work-ing there as an usher in 1948.

Concerned citizens, who recognized the historic value of the theater, rescued the play-house in the late 1970s. Local architects spent three years designing and restoring the struc-ture to its original appearance, preserving the ornate columns and terra-cotta design. Inside, a double flight of stairs leads from the elegant lobby to the balconies of the the-ater. Although the theater's square footage is the same as it was in 1913, the theater today seats only 1,943 guests, compared to its original 2,050 capacity. "My grandparents were much smaller in stature than we are today," notes Morgan.

No historic, cavernous theater would be authentic without a resident ghost, and the Capitol Theatre is no exception. "I feel kind of funny talking about ghosts because I don't actually believe in them, but the evidence is overwhelming," admits Morgan. He tells how security guards make their routine nightly rounds, turning off lights as they

Above: Symphony Hall is the Salt Lake Valley's premier cultural showcase. Built in 1979, the wedge-shaped concert hall is "the most impressive of all (concert halls in the country)," according to Time *magazine. The acoustically acclaimed Symphony Hall was designed by the firm Fowler, Ferguson, Kingston, and Ruben of Salt Lake City. Photo by Stephen R. Smith*

Right: The renowned Utah Symphony performs more than 100 concerts throughout the year. Courtesy, Utah Travel Council

go, only to find the lights back on when they return on their next patrol. "One security guard lasted only one-and-a-half shifts," he laughs. Many people have heard wardrobe crates being mysteriously moved. Upon inspection, the noise stops. "I just talk to him, and tell him—his name's George—that if he doesn't stop the racket, he'll be exorcized. That usually works," says Morgan.

One of the finest theaters in the country is the stately Pioneer Memorial Theatre, built in 1962 on the University of Utah campus. During the theater's 25th anniversary celebration in 1987, the mayor paid tribute to the 3,000 performances that had been given on the theater's stage since the curtains went up in 1962: musicals, Shakespearean classics, comedies, and dramas. Writers extolled the playhouse as the real star on the opening night in 1962. On the night of the inaugural production, the audience arrived an hour before curtain call. Women wore elegant gowns; men wore tuxedoes. It was the social event of the year. Editorials praised "the magnificent building [as] an inspiring symbol of the burgeoning growth and development of the cultural arts throughout our state and nation."

Only steps away from the Capitol Theatre is the acclaimed Symphony Hall, which opened in 1979 after seven years of planning and $12 million in construction. Over the years Symphony Hall has received national recognition.

"The Utah Symphony's warm, responsive Hall in Salt Lake City, built in 1979, is the most impressive of all [concert halls in the country]," reported *Time* magazine in 1983. In 1987 *Time* wrote, "Despite its reputation, Carnegie Hall was not quite as good as . . . newer spaces such as the Philarmonie in Berlin and Symphony Hall in Salt Lake City."

The Utah Symphony consists of 85 musicians and presents more than 250 performances each year. Six international tours have earned the Utah Symphony recognition as one of the finest musical ensembles in the world. Prior to 1979 the orchestra performed in the Tabernacle on Temple Square, home of the Mormon Tabernacle Choir. Now the symphony performs in perhaps the finest musical hall in the country. The three-story glass-and-brick structure is fronted with a 100-foot-long diagonal fountain. The interior is a luxurious combination of brass, gold leaf, natural oak trim, and dark green carpet. Six geometric chandeliers with 18,000 beads of hand-cut crystal imported from Austria and Czechoslovakia adorn the interior.

In 1983 Joseph Silverstein, then concertmaster and assistant conductor of the Boston Symphony Orchestra, was named music director of the Utah Symphony. Maestro Silverstein has increased the orchestra's repertoire, adding classics by Haydn, Tchaikovsky, Mozart, Bach, and twentieth-century composers such as Stravinsky, Sessions, and others.

During 1986-1987 nearly half a million people heard Utah Symphony performances, including 73,000 children. Tchaikovsky's *1812 Overture,* complete with 21 cannons, can be heard in Sunday afternoon performances held in the mountain air of Snowbird during

the summer. Concert-goers in bermudas, sandals, and sunglasses enjoy the music, and many bring coolers, lunches, and jugs of wine.

Hollywood also discovered Utah. Producers used Utah as a location for many Hollywood films, some dating as far back as 1922. Thanks to the Utah Film Commission and Robert Redford's Sundance Institute, Utah is again becoming a creative center for national film and commercial productions. "Of all the 50 states, Utah has the greatest diversity of landscape and amenities within the smallest space," says Leigh Von der Esch, director of the Utah Film Commission.

Robert Redford's Sundance Institute was founded on his Sundance Ski Resort in Provo Canyon in 1980. The institute's intensive workshops and labs assist independent filmmakers, producers, directors, composers, and playwrights in the production and creation of innovative works. "I have always felt the need for artists to have a place where they can try new things, experiment with new ideas and have the freedom to fail," comments Redford. These independent films are shown at the United States Film Festival sponsored by the Sundance Institute. Held each January in the old silver-mining town of Park City, now a world-class ski resort, the festival attracts thousands of moviegoers,

The mountain air and high peaks of Snowbird make it the perfect setting for a summer afternoon with the Utah Symphony. Photo by Stephen R. Smith

amateur film producers, independent filmmakers, and celebrities. The Sundance Institute will introduce its Childrens' Theater in 1990.

Utah's arts organizations survived the nationwide cutbacks in the 1980s mainly because of community support. Salt Lake City's business community rallied to save Ballet West. When the Capitol Theatre was facing closure by Salt Lake County, the county, city, and state made an unprecedented agreement to share the financial responsibilities of operating the theater, which is home to Ballet West, the Utah Opera Company, the Ririe Woodbury Dance Company, and Repertory Dance Theater, perhaps the most traveled modern dance company in the United States.

"Experience shows that downtown vitality is at its peak when planners work with new things around the old rather than starting from scratch," says Roper. Visual arts, theater, dance, music, museums, libraries, nightlife, and history are important factors in deciding where people want to live and work. Quality of life ranks among the leading criteria for U.S. businesses looking to relocate. Because of sophisticated communications systems, businesses can now locate wherever they choose, and cities have begun using their arts to attract businesses. Some 30 American cities have developed arts districts, which have been credited with furthering economic development in blighted areas. Salt Lake doesn't need to look beyond its borders to find excellence in the arts.

One of the biggest problems facing artists is studio space. In St. Paul, Minnesota, city planners polled the city's artists on their needs. The poll revealed that artists needed space in which they could live and work. St. Paul Mayor George Latimer said,

Above: Cynthia Davis is a violin maker and one of two dozen tenants at Artspace, a community that combines living and studio space for working artists. Photo by Stephen R. Smith

Above right: Epicurian delights attract lunch and evening crowds to The New York Club, Market Street Grill, and the Oyster Bar. Photo by Stephen R. Smith

"Artists are a stabilizing force in changing neighborhoods."

Salt Lake City has also realized this to be true. In 1982 Salt Lake's mayor and city council issued a Community Development Block Grant to a group of artists. Their plan was to renovate a turn-of-the-century produce warehouse on the west side of downtown and turn it into living and studio space for artists. "A city needs a cultural district where people know they can go at any time of day to find activity," says Stephen Goldsmith, one of the creators of Artspace. Artspace, a private nonprofit organization, now has 20 studios and houses 27 artists ranging from woodcarvers, dancers, weavers, and photographers to painters, sculptors, and even a violin maker. "Salt Lake is the foremost violin making center in the United States," says Goldsmith, who is a furniture maker, sculptor, urban renovator, and landlord of Artspace. The studios see little turnover, and Artspace is meeting only a portion of the demand. Like many artists, Goldsmith credits Brigham Young and the pioneers with priming the canvas for today's sophisticated arts found in Salt Lake City. "Today Salt Lake's cosmopolitan spirit is nourished through the arts," says Goldsmith.

Not far from Artspace another renovation project is under way in the former gateway district, a once-shabby area west of the central business district. Here graphic arts houses and printers are moving in. Hotels like the old Peery have been restored; on the next block restaurateurs have transformed buildings with peeling plaster and broken windows into some of the city's finest dining establishments: The New York Club, Oyster Bar, Market Street Grill, Cafe Pierpont, Viva La Pasta, and Shenanigans.

The city's appreciation of the visual arts is evident in the number of private art galleries springing up in and around the downtown area. The Salt Lake Arts Center, located between the Salt Palace and Symphony Hall, features visual art exhibits, film series, art classes, lectures, and workshops. The Utah Museum of Fine Arts on the University of Utah campus displays a continuous series of permanent and temporary exhibits, traveling shows, guided tours, lectures, concerts, and films. And the annual Utah Arts Festival has grown so large over the last few years that organizers have had to move it from West Temple Street to the open grounds of the Triad Center. Here, artists from around the country display and sell their paintings, jewelry, carvings, and woven goods to throngs of people. Outdoor concerts, culinary delicacies, and children's programs offer something for everyone.

Salt Lake City also has plenty of children's arts programs, including Children's Dance Theater, the Children's Museum of Utah, and Lallapalooza, an organization that co-sponsors workshops all over the city, including "gardening and art" with the state arboretum. At the Children's Museum, youngsters let their curiosity lead them through imaginative play. Here they can pilot a 727 jet, experiment with frozen bubbles, implant an artificial heart, anchor a "KIDS" television news program, or learn about the value of money in the Kids' First Bank or the Kids' Corner Market.

Workers, shoppers, and tourists in downtown Salt Lake City can enjoy the Brown Bag Concert Series, which offers 57 fully produced, free, noontime concerts in Crossroads Plaza, ZCMI Center, mini-parks, or outdoor plazas.

THE LOCAL SPORTS SCENE

People can attend an NBA basketball game as well as a ballet performance. Many Utahns have become instant fans of the Utah Jazz, Utah's NBA basketball team. During the 1987-1988 season the Jazz played to 40 sellout crowds out of a total of 41 home games and took the world-champion Los Angeles Lakers to the limit in their best-of-seven series in the conference semifinals. "Ya gotta love it baby!" says "Hot" Rod Hundley, the voice of the Utah Jazz. The Jazz have become so popular that they have outgrown the 12,500-seat Salt Palace, and plans are under way to begin construction of another arena.

In the shadow of the Jazz, the winning Golden Eagles hockey team—International Hockey League champions in many seasons—still has plenty of loyal fans. The Salt Lake Trappers of baseball's Pioneer League also attract thousands of fans to Derks Field each summer in hopes of repeating its all-time baseball record of 29 consecutive wins in 1987.

The Utah Jazz, the state's NBA basketball team, plays to sellout crowds. Courtesy, Utah Travel Council

In 1988 the Salt Palace served as the host location for the U.S. Olympic Gymnastics Trials. The Salt Palace is Salt Lake's major civic auditorium, hosting Utah Jazz NBA games, Golden Eagles hockey games, and many concerts and conventions. Photo by Stephen R. Smith

Most Salt Lake football fans get caught up in the rivalry between the University of Utah Utes and the Brigham Young University Cougars. Even downtown business executives are loyal to either the red or the blue. It's not uncommon for scores, plays, and running and passing records to dominate preliminary discussions at board meetings and luncheons.

Salt Lake residents also are sophisticated gymnastics fans, having cheered the University of Utah women's team to six consecutive NCAA championships. In 1988 the U.S. Olympic Gymnastics Trials were held at the Salt Palace, drawing thousands to watch athletes qualify for the 1988 Olympic Games in Seoul, South Korea.

RESTAURANTS AND RESORTS

Dining pleasures in Salt Lake are unlimited. Perhaps Salt Lake's finest restaurant, La Caille at Quail Run is nestled in the woods in Sandy near the entrance to Little Cottonwood Canyon. La Caille at Quail Run serves French cuisine in a country garden setting while geese, ducks, and rabbits roam the grounds surrounding the chateau and the cobblestone drive.

Up in the nearby canyons the ski and summer resorts of Alta, Snowbird, Brighton,

The Santa Fe in Emigration Canyon is one of the area's newest and most popular restaurants. Photo by Stephen R. Smith

and Solitude have intimate, quiet restaurants serving exquisite dinners and Sunday brunches in the mountain air. In addition to its fine restaurants, Snowbird hosts concerts, dance workshops, cultural festivals, painting and photography courses, and nature workshops. "My underlying dream for Snowbird involves the creation of a man-made environment that matches the inspirational nature of the physical environment—a place for people that contributes to the growth of the total individual," says Richard Bass, Snowbird owner. Indeed, the Snowbird environment caters to the body, mind, and soul. After a full day of skiing, people can unwind in the new, luxurious Cliff Spa. The spa features the latest unique health and fitness services to "nurture the body, relax the mind, and refresh the spirit." Visitors can treat themselves to hydro massages, herbal wraps, and facials or work out in the swimming pools or weight room.

Even the smaller canyons rimming the edge of town offer canyon dining experiences. Log Haven in Millcreek Canyon, Ruth's diner, and Santa Fe in Emigration Canyon are three local restaurant favorites. The Santa Fe recently opened in a remodeled building. Spanish-style mosaic tile and glass adorn the entrance. Large clay pots, sofas with patterned pillows, and watercolor paintings of muted greens, corals, and blues depict southwestern scenes of panoramic sunsets, desert bones, and phantom coyotes. A 10-foot handcarved snake curves along the top of the fireplace. In the dining room, small cactus plants accent each table. The menu features spicy, southwestern dishes like fried chilies, blackened snapper, and sesame chicken.

Park City, less than one half hour east of Salt Lake, is a historic silver-mining town nestled in the mountains. Today it is known mostly as a ski resort village and is home to the U.S. Ski Team, but Salt Lakers like to go there for other reasons. Fine restaurants, boutiques, gift shops, and art galleries line the city's steep, historic Main Street. Quaint turn-of-the-century houses, painted in soft yellow, white, and blue, are dwarfed by acres of dark brown and gray condominiums and resort hotels. In August thousands of Utahns crowd Main Street during the Park City Arts Festival. In September they converge on the dewy Park Meadows Golf Course in the early chill of dawn to see Autumn Aloft, the annual hot-air balloon festival. Winter, of course, brings skiers from the world over to the powdery, steep slopes of nearby Deer Valley, Park City, and Park West ski resorts.

A PLACE TO CALL HOME

Utahns are proud of their ancestors and celebrate and honor their pioneer heritage in a number of annual festivals. Perhaps the most popular festival is the Days of '47 Parade, celebrated on July 24, a state holiday. This is the second largest festival of its kind in the country and features more than 200 hand-made floats, a rodeo, and a fireworks display. Another annual festival, Neighbor Fair, attracts an estimated 150,000 people and honors the hundreds of nonprofit organizations that exist in the valley.

Utah is host to other festivals, including the Festival of the American West in Logan, Springdale's Southern Utah Folklife Festival near Zion National Park, and the Pageant of the Arts in Springville. The annual Utah Shakespearean Festival in Cedar City has become nationally acclaimed. Many Salt Lakers attend one of several performances presented from July through August, whether they are Shakespeare buffs or not. Some avid playgoers attend all three productions, while others simply go to indulge in the medieval banquet, afternoon teas, or costume calvacades. Organizers had to extend the festival in 1988 to accommodate more guests and will likely have to do so again in 1989. By the close of the 1988 festival, the box office had already sold thousands of tickets for 1989.

Other festivals honor the old-world traditions of many of Utah's communities. Through songs, stories, dances, music, and foods, Utahns can celebrate the rich heritage of other cultures. And for many first- and second-generation Americans these festivals provide a link between their new life in America and the time-honored traditions of their homelands. Ethnic festivals allow all residents to be a part of the present, without losing the past.

For example, the Greek Festival, held each September at the Hellenic Memorial Cultural Center adjacent to Holy Trinity Greek Orthodox Cathedral, is one of the nation's

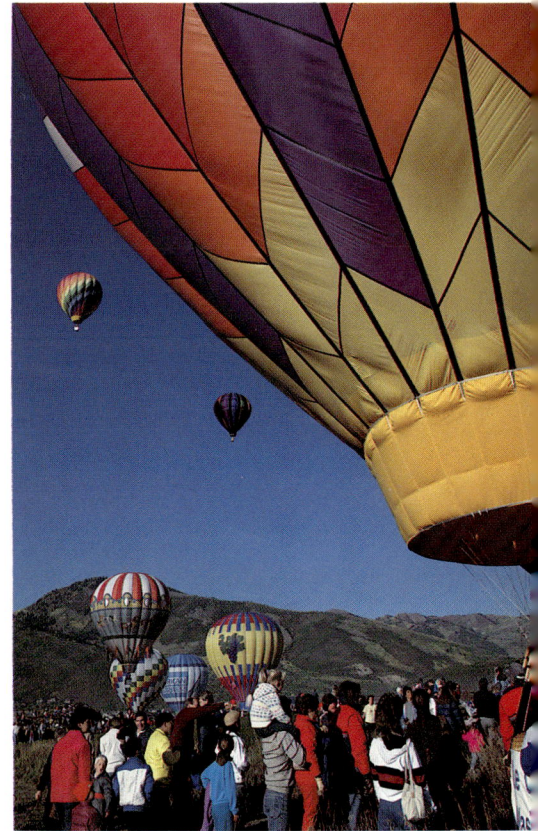

The annual Autumn Aloft balloon festival in Park City attracts balloonists from around the country. Thousands of spectators rise early to witness this festival of color. Photo by Stephen R. Smith

Besides being home to a major ski resort, the village of Park City offers many year-round attractions. Courtesy, Utah Travel Council

Besides being home to a major ski resort, the village of Park City offers many year-round attractions. Courtesy, Utah Travel Council

largest ethnic celebrations. Residents throng to this annual festival to sample baklava, cheeses, pastas, herbs, olives, and other rich delicacies, and perhaps dance a few numbers.

At Oktoberfest, held every autumn at Snowbird, crowds pack the pavilion for a taste of Bratwurst and beer, stomp to the lively sound of the accordion, and learn a folk dance or two. At Swiss Days in Midway, valley residents often return with a pair of lederhosen.

The annual Living Traditions Festival celebrates the breadth and diversity of Salt Lake's cultural heritage and includes music, dance, crafts, and foods from Salt Lake's many culturally distinct communities.

Christmas in Salt Lake City brings people from all parts of the West to the downtown area. The trees along Main and State streets and the Temple grounds are decorated with millions of twinkling lights, illuminating Temple Square like a storybook scene. Shoppers stroll the sidewalks or ride huddled together in horse-drawn carriages, humming carols. Choral groups, pipers, and bell-ringers fill the malls and streets with music, spreading good cheer. At the Salt Palace, families attend the annual Festival of Trees. Each one-of-a-kind Christmas tree is decorated with a theme or craft, such as dolls, bears, ribbons, wreaths, or bows. The trees are later auctioned off, and the proceeds go to Primary Children's Medical Center.

These cultural events shape the lives of all Salt Lakers. The Mormon Church

plays a significant role in the well-being of the community. More than 2,500 people—Mormons and non-Mormons—visit the Family History Library of the Church of Jesus Christ of Latter-day Saints each day to research their family ancestry. This library is the world's largest collection of family history records and is open to the general public at no charge.

"Utah will always be linked with its LDS heritage, as many areas of this country are linked with specific religious groups, such as Boston and Roman Catholicism, or Virginia and the Anglican heritage of its founders," writes the Right Reverend George E. Bates, Episcopal Bishop of Utah. "As the various areas of our country have moved into the twentieth century, they have recognized the need to embrace the cultural, ethnic and religious variety of their citizens."

When Dino Georgalas, manager of the Marriott Hotel, first came to Salt Lake, he was amazed to find squash courts throughout the Salt Lake Valley. He grew up in Egypt and England and played the game in many countries, but when he moved to the East Coast, squash courts were available only to the rich and famous. Even "in Los Angeles, it's rare to find squash courts," he comments. He attributes this evidence of Salt Lake's worldly culture to the Mormon Church. "Missionaries bring cultures of the world back to Utah. And more Utahns speak foreign languages than in other states." He enjoys honoring the various cultures of his employees by attaching flags to their name tags representing their native countries.

Neighbor Fair is part of the annual Days of '47 festival, the state holiday on July 24th which commemorates the day in 1847 when the Mormon pioneers entered the Salt Lake Valley. Photo by Stephen R. Smith

Salt Lakers make good neighbors, especially in tough times. During the notorious floods of 1983, the community pulled together to save the city and each other. Most residents still recall those long days and nights.

In May 1983 sudden hot temperatures melted the deep snowpacks in the mountains above the valley floor. The potential of a major flood dominated newscasts around the country. Former mayor Ted Wilson quickly consulted engineers and decided to divert the runoff of City Creek through town, transforming State Street into a wide, eight-inch-deep river. Thousands of sandbags needed to be filled immediately to bank and contain the waters. Thousands of men, women, and children turned out to fill, load, and distribute the sandbags. People volunteered throughout the nights, working shoulder to shoulder with strangers, united in their effort to prevent the potential disaster. Neighbors took turns bailing out each other's basements and cleaning up farmlands, efforts that were heralded across the nation.

Residents still pull together today in the face of economic strife. On the west side of town, hundreds—the numbers change daily—of homeless men, women, and families live in temporary, crowded shelters and trailers or under freeway overpasses. Instead of ignoring the problem or just living with it, Salt Lake's residents, businesses, and religious groups have chosen to deal with the situation in a positive way. Citizens have responded to the mayor's request for donations to construct new shelters. These shelters, opened in November 1988, can house 110 people in a family unit and 237 in a single men's unit. Another unit for single women will be added later. Support services are also available near the facility and include an employment office, soup kitchen, transportation facilities, and a thrift store.

Salt Lakers and visitors alike appreciate the city's cleanliness and wholesome life-style. "As I walked, I noticed that Salt Lake seems less chaotic (both physically and socially) than most American cities. Many of its inhabitants, wrapped in an aura of spiritual certitude and traditional values, go about their business with the cheerful confidence of people who know exactly how they fit into the scheme of things," commented a writer in *National Geographic Traveler* magazine.

Salt Lake is a community that has had to develop its own assets. Now, like other communities large and small, it is looking outside its borders for investors, visitors, developers, and corporate relocators to broaden its tax base and raise the standard of living. The city, though, already offers plenty of livable qualities. "Quality of life means different things to different individuals," says Mayor DePaulis. "It might be a job, an education, affordable housing, the natural environment, community involvement. Quality of life is whatever makes you happy."

"Salt Lake is one of America's last large cities where you can live downtown and enjoy safety, cleanliness, privacy and quiet," beams Georgalis. "It's the Switzerland of America, with the best natural beauty seen anywhere. When my friends asked why I'd want to leave Los Angeles, especially after living in Egypt, England, Switzerland, and Canada, to move to Salt Lake City, I said 'Here I enjoy qualities that money can't buy.'"

It isn't by accident that so many of Salt Lake's finer points center around its heritage, its unique composition of people, and the importance of the family. "Our community has grown to appreciate these finer qualities of our lives," says June Morris. "I've been all over the world, and I'm always happy to come back to Utah. I wouldn't live anywhere else." Her response to people looking to move to Salt Lake is "Come. It's not threatening at all. People who've moved here hate to leave for the most part."

City leaders hope the R/UDAT study will encourage more citizens to get involved in building their community's future. Ambivalence bears apathy at best, complaining at worst, and gives free rein to special-interest groups. The R/UDAT's message is that all residents need to express themselves and stumble through the inevitable frustrations but still keep moving ahead. Citizens should speak their minds and share their visions. Perhaps Henry David Thoreau expressed it best in his book *Walden*:

I know of no more encouraging fact than the unquestionable ability of man to elevate his life by a conscious endeavor. It is something to be able to paint a picture, or carve a statue, and so to make a few objects beautiful; but it is far more glorious to carve and paint the very atmosphere and medium through which we look, which morally we can do. To affect the quality of the day, that is the highest of arts.

The family is an important aspect of life in Salt Lake and the area provides many occasions for family-oriented events. This young family enjoys some winter vacation time in nearby Alta. Photo by James W. Kay

CHAPTER EIGHT

THE GREAT OUTDOORS

When recently surveyed about the benefits of doing business in Salt Lake, 90 percent of 600 chief executive officers said they most enjoy Utah for its scenic beauty and its recreational activities. The unparalleled natural beauty of the rugged peaks, alpine meadows, and desert expanses that surround the Salt Lake Valley plays a significant role in the lives of most Salt Lakers. Whether it is a businessman who takes his children to a favorite fishing hole after work, a student who heads for the ski slopes after class, a couple that spends the weekend backpacking in one of Utah's five national parks, or a woman who watches the sun rise over the Wasatch pinnacles, all find joy in the simple pleasures nature offers.

To paraphrase the late Edward Abbey, a controversial writer and conservationist, "perhaps not everyone chooses to adventure in the out-of-doors, but they are comforted just knowing it's out there, untamed, pristine, balanced. They may never climb Mount McKinley or Everest, but they are grateful they exist."

Everyone enjoys the bounty nature provides—the clean air and pure water. Everyone enjoys a family picnic in the canyons or a quiet morning walk to a mountain lake, but only a few people are committed to preserving these finer qualities of life.

Situated in a mountain basin, Salt Lake City has four distinct seasons. January temperatures average 37 degrees, but August temperatures often reach the high 90s. Chilling rains around Halloween send the temperatures falling to the low 60s. During the long winters it snows heavily in the mountains, but neighborhoods along the foothills often get only between one to two feet of snow. In the valley, snow may fall for days—or not at all. Road crews respond immediately after the snowstorms, usually clearing the roads overnight.

Spring in Salt Lake City brings long-awaited sunshine, a flurry of tulips and flowering dogwood, fruit, and olive trees, cascading canyon streams, and, yes, more skiing. Spring skiing at Utah's resorts usually extends into June, since Memorial Day snowstorms are common. Snow clings to the mountain cliffs well into July. Summers mean arts and cultural festivals, barbecues, picnicking, fishing, camping, and hiking in the surrounding mountains. In autumn, hillsides are painted a deep red. Canyons of aspens, cottonwoods, fir trees, and scrub oak become a palet of glistening golds, yellows, burnt oranges, reds, and dark greens. This is the season when even the most sedentary folk flock up into the canyons to witness the glorious display of color in nature. When it is too hot in the valley, residents retreat to the high mountains and cool canyons of the Wasatch or the Uinta, two hours east.

For decades the canyons that cut through the Wasatch range have served as playground, watershed, and wilderness escape for city dwellers. Nearly 80 percent of the canyons are public lands, managed by the National Forest Service. Traditionally, the Wasatch forest has been one of the most heavily used forests in the entire United States. This popularity is perhaps due to the promotional literature, which describes the 11 ski resorts within an hour's drive of Salt Lake City: Brighton and Solitude in Big Cottonwood Canyon; Alta and Snowbird in adjacent Little Cottonwood Canyon; Deer Valley, Park City, and Park West near Park City; Sundance near Provo; Snow Basin and Powder Mountain near Ogden.

Tourism officials are proud of Utah's ski industry, claiming it contributes millions to the economy in seasonal jobs and services. The state's new license plates also advertise the commodity: "Ski Utah. The Greatest Snow on Earth." The

Some of the best skiing in the country can be experienced in the Salt Lake Valley area. Photo by James W. Kay

Right: With a remarkable season that runs from November through May each year, the ski industry provides a major contribution to the state's economy. Courtesy, Utah Travel Council

Below: Warm summer months offer canoeing enthusiasts the opportunity to glide through the waters of Silver Lake Flat Reservoir in American Fork Canyon. Photo by Stephen R. Smith

Above: Wildflowers in the Uinta Mountains barely have time to blossom during the spring and summer months, before the snow begins to fall once again. Photo by Stephen R. Smith

Left: Mountainsides of golden aspens, red oak, orange maples, and emerald firs surround the Salt Lake Valley in spectacular color every autumn. Photo by Stephen R. Smith

resorts operate from November through May and make most of their money on holiday weekends. For this reason, most resorts are open by Thanksgiving, provided nature cooperates and sends lots of snow.

Perhaps the most famous Salt Lake ski resort is Snowbird. Its 1,900 skiable acres (3,100 feet from top to bottom) have terrain of varied difficulties. Its 125-person aerial tramway and 500 inches of annual snowfall make it one of the most popular ski resorts in the West. The resort's ski center, full-service hotels, and time-share accommodations are constructed of concrete and glass to "complement the surrounding environment." The new, $80-million Cliff Lodge features a 10-story atrium which is walled in glass. From the sunken restaurant, patrons gaze up to 11,000-foot peaks. "You always feel like you're outside. Even the bathrooms have windows with a view," says a Snowbird spokesman. Snowbird's sole owner, Dick Bass, has climbed the highest peaks on the world's seven continents. He designed the Cliff Lodge while holed up in a tent on one of his expeditions. His love of climbing prompted him to open a mountaineering center at Snowbird and attach a technical climbing wall to the expanded Cliff Lodge. In 1988, international climbing competitions were held at Snowbird, the first competitions of this kind in North America.

Alta, a few miles from Snowbird, is a historic gold-mining town. Alta is a favorite among local skiers because of its steep slopes of deep powder. At the top of Big Cottonwood Canyon sits Brighton, a quaint village of Swiss-looking cabins, a general store, a lodge, and a ski shop, all clustered in a mountain bowl. The terrain of Brighton's four lifts varies from beginner to expert. Solitude, located down the canyon, is popular with local skiers for its rustic simplicity and shorter lift lines. The Utah Ski Association says ski resorts are a great use of land. "You have a lot of people enjoying themselves on a very limited amount of land, leaving the majority of the land to the people who don't

Snowbird is nestled in a breath-taking mountain valley. Photo by James W. Kay

want to be around a lot of people," says Bob Bailey, executive director.

Not everyone gets their thrills by riding to the top of a mountain, standing on two boards, and careening down the face at breakneck speed, repeating the ritual several times in the course of the day. A growing number of Salt Lakers are turning to cross-country, or Nordic, skiing because of its cardiovascular benefits and less expensive equipment, and because of the freedom it allows to explore the backcountry or groomed courses.

Salt Lake City lies a half an hour away from three wilderness areas: Mount Olympus, Lone Peak, and Twin Peaks. These are some of the most remote regions of the Wasatch. Peaks, ridges, forests, talus slopes, lakes, and meadows extend as far as the eye can see. Here, hawks circle overhead, mountain goats graze from precarious heights, and waterfalls spill through crags and over the granite faces of Little Cottonwood. Early morning hikers can startle cottontail rabbits and mule deer from their beds in the alpine meadows. Acres of white columbine, lavender aspen daisies, silvery lupines, purple larkspur, rosy paintbrush, and the magenta petals of the parry primrose sweep over the landscape. Running through the Wasatch are the canyons of City Creek, Red Butte, Emigration, East, Parleys, Millcreek, and Big and Little Cottonwood.

Spring floods and summer droughts have heightened concern over preservation of the canyons. For the first time, Salt Lake County has initiated a master plan to protect the canyons' critical water supply and to help balance development and commercial interests against conservation concerns. These canyons serve as special places where local residents can experience the beauties of nature.

Nearly all of the canyons are slowly being transformed into suburbs. Each year more houses spring up along the foothills. Despite the master plan's moratorium on construction, the county recently approved expansion plans for Solitude. The resort wants to

Little Cottonwood Canyon is one of many ancient, glacier-carved canyons that rim the Salt Lake Valley. It is home to Snowbird and Alta ski resorts, and is a critical source of water for valley residents and businesses. Photo by Stephen R. Smith

add two new lifts—increasing skier capacity by 40 percent—build condominiums, and remodel existing buildings in a European style. Brighton wants similar approval. In appealing the county's decision, concerned citizens noted that the increase in skier capacity "will have substantial adverse impact on the many types of winter recreational uses of Big Cottonwood Canyon."

Yet more people visit the canyons to hike and picnic than to ski. In fact, more than 3.5 million people visit the canyons each year. The Wasatch Front canyons receive more visitors than does Yellowstone National Park. Picnic areas line the streams in Big Cottonwood, Little Cottonwood, and Millcreek canyons, and more and more young people and families are hiking to high lakes and glacially carved bowls. People enjoy going to the mountains, for whatever reason. For most folks a day in the canyons is a needed escape from the noise and traffic of the city. But when conflicts follow you into the wilderness, it degrades the backcountry experience. "We have to work out a proper balance. I don't think the state's pressure for growth and the pressures of population should despoil those canyons for short-term economic gain," says Ann Wechslar of the Sierra Club.

An organization called Save Our Canyons is opposed to further commercial construction in the canyons, saying the ski resorts are quickly overtaking lands used by backpackers, naturalists, and cross-country skiers. Furthermore, ski resort development may not be a wise investment when cross-country skiing is growing much more rapidly. The organization believes cross-country skiing will remain a popular sport.

"Sure, the ski resorts are attractions to a certain point," says Alexis Kelner of Save Our Canyons. "But the state promotes them as being one half hour away from the Salt Lake City center, so why do we need another major city in the middle of the mountains?" Those dedicated to preserving Utah's forests and canyons support limited growth of the ski resorts and believe development should be restricted to ski area boundaries.

Alta, the only incorporated town in the canyons, has imposed a building moratorium of its own. The community implemented zoning regulations to control growth when the city first incorporated. "I'm very proud of our people. They've opted for life-style and preservation rather than a quick buck," comments Alta Mayor William H. Levitt.

The steep, winding highways leading up to the resorts—only single lanes in both directions—cannot accommodate the increasing traffic that bigger ski resorts or the Winter Olympics would attract. Several proposed solutions to the traffic problem have been suggested, including better use of existing bus transportation; a massive $400-million

Millcreek Canyon is a popular picnicking, hiking, and fishing area in the summer and a favorite cross-county ski path during the winter months. Photo by Stephen R. Smith

supertunnel that would connect the Salt Lake Valley with Snowbird, Alta, Brighton, and Park City; an aerial tramway interconnect system; or monorails that would allow skiers to park at the canyon base and ride to their favorite resort. Tourism and ski industry officials and consultants say the interconnect would create the largest ski complex outside of Europe. In any case, the master plan prohibits the resorts from expanding parking space. "There's plenty of open spaces for people in our mountains," says Ted Wilson, former mayor and active outdoorsman. "There's just no room for more cars."

The Wasatch Mountain Club suggests other solutions to the canyon car congestion, such as stopping cars with fewer than three riders, requiring ski areas to reduce lift fees for bus riders, and collecting parking fees at ski areas.

Many citizens are against further canyon development. Some say that prohibiting cars in the canyons will enhance mountain visits and that the mountain setting is more picturesque without cars and buses. Others agree that shipping more people up the canyon en masse is not in Salt Lake's best interest. In any case, it is important that valley residents be a part of the planning process and play an active role in preserving Salt Lake's canyons.

Though the popularity of downhill skiing continues to develop at a rapid pace, there is still plenty of pristine wilderness surrounding Salt Lake to accommodate all outdoor activities. Photo by James W. Kay

WATER

Utah has the largest inland lake and yet, surprisingly, is the second-driest state in the nation. Although it is considered an arid region, Salt Lake City has not imposed—at least in recent history—any water rations or usage restrictions during dry seasons. Lawns are green, sidewalks and gutters are washed clean, and cars are spotless, despite constant water concerns.

Salt Lake homes receive about half of their water supply from neighboring canyons. As reported on KTVX news, protecting that water quality is a challenge when snow is plowed off ski resort parking lots into the streams. Snowmelts are highest in the late spring before water consumption peaks in late summer. Surrounding reservoirs such as Deer Creek, Mountain Dell, and the new Jordanelle (under construction) provide the remaining half of the valley's water supply, and are popular weekend camping, fishing, waterskiing, and sailing spots as well.

In 1983 sudden hot temperatures quickly melted the deep snowpack in the high mountains, sending the water raging down the canyons and into the city where the streams flowed into the Jordan River, which empties into the Great Salt Lake. The lake climbed to record levels, flooding farms, burying railroad tracks, closing lakefront mineral businesses and the sailboat marina, and lapping up the newly rebuilt and reopened Saltair beach resort at Great Salt Lake State Park. The famous lake, legendary for its salt content, lost much of its salinity over the year that followed. Early projections predicted the same would happen in 1984. To avert another crisis, Governor Bangerter authorized the installation of $60-million pumps to drain some of the lake's water into the western desert

Right: High in the Uinta Mountains east of the Salt Lake Valley, Three Lakes Divide becomes the headwaters for the Provo and Weber rivers, two major water sources feeding the lower valleys. Photo by Stephen R. Smith

Facing page, top: Though Utah may be considered an arid state by most, members of the Great Salt Lake Yacht Club find plenty of opportunities to participate in their favorite pastime of racing around the salty inland sea. Photo by Stephen R. Smith

where it would evaporate. The lake's level is now under control and regaining its salinity. The floods never came. Instead, Salt Lake has seen several years of drought.

WEEKEND GETAWAYS

Two hours east of the valley, the Uinta Mountains have some of the state's highest peaks and most scenic panoramas found anywhere in the area. Hundreds of lakes dot the area, from roadside fishing ponds to high, secluded alpine lakes filled with bone-chilling water and rainbow trout. Wildflowers of all colors burst through thawing meadows when the snow finally recedes in July, and have just enough time to bloom before the snow flies again in late September. Yellow marmots (rock chucks) whistle while sunning themselves on giant boulders, where they make their home in the rockslide of a talus slope. Even during the hot days of August in the valley, the temperatures at these 11,000-foot elevations are cool and the air is bracing. Backpackers can set their watches by the afternoon thunderstorms that roll overhead.

Pleasant spring, autumn, and winter temperatures in the desert are perfect for exploring the southern part of the state. Salt Lake is within a day's drive of 10 national parks. Five of them are four to six hours away in southern Utah. Arches National Park, in southeastern Utah, has the world's greatest concentration of natural arches. Throughout the ages combined forces of wind, ice, and water have sculpted these delicate forms in the soft sandstone. This natural gallery of sculpted camels, elephants, eagles, fins, and faces gaze out on the petrified sand dunes, squatty junipers, prickly pear, primrose, and Indian paintbrush, oblivious to the presence of visitors.

Further south lies Deadhorse Point State Park and Canyonlands National Park, a labyrinth of deep canyons, high plateaus, and flat mesas. Most of the park is inaccessible without good hiking boots and a daypack. The Colorado River and its twin river, the Green, flow through here. Visitors can see the tiny ribbons of water from overlooks hundreds of feet up, or by running the infamous rivers and their rapids on a raft. Those looking for a slower, scenic ride can canoe the wide, serpentine stretches of the Green.

To the southwest of Salt Lake, peach and apple orchards line the main road that

winds through Capitol Reef National Park. The park is named for a rock dome that resembles the nation's capitol. Orange domes the size of buildings rise from the earth's surface, separated by narrow, rocky washes that in a rainstorm can fill with water in a second.

Further south is Bryce Canyon National Park. At Bryce, visitors walk through stands of pine trees to the edge of the canyon that looms abruptly ahead. Spires of rocks colored orange, coral, salmon, and buff from mineral runoffs rise from the canyon floor like needles in a pin cushion. Trails lead straight down along the maze of columns into the valley floor, leading the hiker or cross-country skier past towering pines growing from the crumbling rock. Although Bryce Canyon is in the desert, some points in the park sit at 9,000 feet above sea level.

Next is Zion National Park. Here, the visitor looks up at the geological features from the canyon floor. A short, steep hike leads to Weeping Rock. Hardier hikers may want to attempt Angel's Landing, the Emerald Pools, or Zion Narrows.

At any of these national parks, visitors can camp in the limited campground facilities or stay in comfortable lodges and motels nearby. To really see the areas, however, they should park the car and walk a ways.

Even parts of the desert that have not been designated as national parks are becoming popular weekend exploration grounds for adventurous Salt Lakers. The San Rafael Swell is a popular hiking spot and is only three hours south of the city. Uplifted from

Capitol Reef National Park is only four hours south of Salt Lake City and is one of Utah's five national parks. Photo by Stephen R. Smith

the surrounding land, the Swell's dome has been eroded away, leaving the two sides. Below the dome are small chutes that twist around hollows, washes, and shoulder-wide narrows lined with cliffs that rise hundreds of feet overhead.

The Grand Canyon is a short distance across the Arizona border, and Salt Lake City is a day's drive from the nation's oldest national park, Yellowstone National Park in Wyoming, and the nation's newest national park, Great Basin National Park in Nevada. Grand Teton National Park, only five hours away, is home to one of the most breathtaking and rugged mountain ranges in the world.

Although residents want new industry and the jobs they bring, they do not necessarily want growth. Most residents hope for a balance between a clean environment and economic security. "Elsewhere I've lived, life is so complicated; You have lines everywhere. Here, we're five minutes from nature, and it's not a hectic life-style," asserts June Morris. "This is what sets Salt Lake apart from the rest. Most of us who live here, whether we realize it or not, would like life to stay like this. But we realize our children will need jobs. Growth is necessary, but I hope it can be managed to preserve and balance the qualities we cherish about living here." That can be done, she believes, but only if people constantly remind Salt Lake's leaders that Utahns must preserve the state's natural treasures.

Organizations such as Tracy Aviary in Salt Lake, the Museum of Natural History and the State Arboretum at the University of Utah, Hansen Planetarium, and Hogle Zoo are educating adults and children about the environment, especially the importance of protecting wildlife, birds, trees, and wildflowers. Field trips, workshops, demonstrations, bird shows, nature fairs, hawk watches, naturalist weekends, and summer geology and biology courses for schoolchildren are teaching the value of the greater, natural world that exists outside the home, school, neighborhood, and boardroom.

Tracy Aviary is one of only two such facilities in the country and has the seventh largest bird collection in North America, with 230 different species and around 1,000 individual birds. Out of the 230 different species, about 80 of them are native to Utah and can be seen in anyone's backyard. Salt Lake is on the Pacific flyway, a major migration route for wild ducks, great blue herons, snowy egrets, and Canada geese. Utah is also one of the largest wintering areas for the bald eagle.

Apparently residents are eager to learn. In the aviary's second year, the attendance at its many educational programs and nature walks more than doubled. Although the aviary cares for 50 percent more birds now than it did a few years ago, it has not received any additional state funding. Corporate and individual sponsorship and donated advertising have helped the skeleton staff to care, feed, and breed birds, as well as educate the public.

"People like the wilds, and the boundless feelings of nature," says Mark Stackhouse, education coordinator at the aviary. "These qualities make Salt Lake a nice place to live, which generates growth. But this same growth threatens the qualities that make it liveable in the first place. We don't want to kill the goose that laid our golden eggs."

Henry D. Thoreau once wrote "A man is rich in proportion to the number of things which he can afford to let alone." What Utahns have to consider is this: would they rather live in Chicago, New York, or Los Angeles? Do they want to contend with the problems these cities have? The advice of the R/UDAT study is sound: "Just say no to some projects."

Salt Lake is a city of balance and harmony, the cosmopolitan heart of a much larger, natural environment—centered in the West, surrounded by rural spaces, rooted in tradition, and inspired by nature. Writers and artists have long been aware of this community's natural beauty. And now city fathers, community leaders, and residents are beginning to share that appreciation. For Salt Lake to continue to have both economic prosperity and a desirable quality of life, each individual and business must do their part. Some sacrifices will have to be made.

Salt Lake's environmental chiefs don't think a clean environment and economic development are mutually exclusive goals.

"Everyone wants clean industry," says Jim Brande, director of the Salt Lake

An excellent weekend getaway can be had by camping high in the mountains and canyons that surround the valley. Photo by James W. Kay

County Health Department's Air Pollution Bureau. Kent Minor, the department's Water Quality and Hazardous Waste director, agrees. "There's no conflict with industry and the environment," says Minor. "Most act responsibly. Everyone wants a clean environment."

Salt Lake is making great strides by complying with federal pollution laws and enacting other innovative programs to clean up its environment. BP Minerals has initiated several pollution control ordinances. Utah offers tax incentives to companies that install pollution control devices. Car owners must have emissions tests performed on their vehicles to control the levels of carbon monoxide in the city. Other companies are excavating or containing or removing hazardous waste sites. A water treatment plant in North Salt Lake will soon be transformed into a wetland bird refuge with a nature trail. And the city's cloud seeding experiments will clear the valley of dirty, stagnant air during cold winter inversions.

In one area, however, Salt Lake is not as progressive as many other western cities: recycling. The very fact that Salt Lake currently doesn't have a waste disposal problem is something of a problem in itself. Still, Salt Lake is in a position to learn from other cities' errors. "Recycling comes about when you have no more room to dispose of garbage. Perhaps Salt Lake will see a bottle recycling bill within the next five years to extend the life of the county landfill," says Minor.

Salt Lake will have become a true success in the new century if, around it, streams still run; aspens, pines, and cottonwoods still live out their ancient cycles; and elk, antelope, and buffalo still roam. The opportunities that natural settings provide—reading a book by a stream, watching a deer and her fawn step from the forest, walking by a rainbow of wildflowers, sailing toward the sunset on the Great Salt Lake—are important parts of life in the Salt Lake Valley. Although Salt Lakers must make way for economic growth, they must proceed cautiously and wisely.

Children learn about the value of nature at the Tracy Aviary. Photo by Stephen R. Smith

A lone rock climber makes his way up a steep wall at Canyonlands National Park. Photo by Mark Gibson

Salt Lake is alive with economic and cultural opportunities and enjoys a peaceful and optimistic attitude about the future. Photo by Steve Greenwood

PART TWO

SALT LAKE VALLEY'S ENTERPRISES

CHAPTER NINE

NETWORKS

T he Salt Lake Valley's energy, communication, and transportation providers keep products, information, and power circulating within the area.

Delta Air Lines, 126-127

Questar Corporation, 128-129

Intermountain Power Agency, 130-131

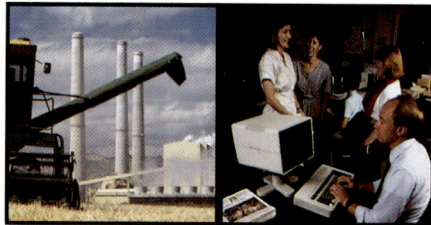

Utah Power & Light Company, 132

Deseret News, 133

Bonneville International Corporation, 134-135

KUTR-AM Radio, 136-137

Photo by Stephen R. Smith

DELTA AIR LINES

Delta Air Lines, which merged with Western Air Lines April 1, 1987, is America's oldest existing air carrier. The company dates back to Western's first flight, with bags of airmail, from Los Angeles to Salt Lake City on April 17, 1926.

Soon after those first airmail flights began, Western took to the skies with paying passengers, becoming the nation's first airline. That same year a company called Huff Daland Dusters was providing crop dusting—a business that began in 1924—to farmers from a base in Monroe, Louisiana. Under the leadership of C.E. Woolman, Huff Daland became Delta Air Service and inaugurated passenger service on June 17, 1929.

From these aviation pioneers has grown the Delta Air Lines of today—a carrier that provides Salt Lake City with more than 150 flights daily. Delta operates its third-largest hub complex in Salt Lake City, providing nonstop and one-stop service to 72 cities. This service provides a meaningful boost for area business prospects, as Salt Lake City enjoys a higher level of flight services than most U.S. cities of comparable size.

Delta's Salt Lake City hub is steadily growing in popularity as traffic increases beyond the previous levels established separately by Delta and Western. For example, in December 1987 Delta boarded nearly 8 percent more passengers than Delta and Western combined exactly one year earlier.

Following the April 1, 1987, merger, Delta made a strong commitment to Salt Lake City by opening a new marketing/training facility—including a reservations center to handle calls from throughout the western United States—at a cost of more than $9 million. More than 20,000 reservations calls are received each day by Delta personnel at the center. By the end of 1988 Delta had added some 500 additional personnel to its Salt Lake City work force in order to staff the new facility. In addition, subscribers to Delta's DATAS II/DeltaStar reservation system receive training at the Delta Salt Lake City Automation Training Complex, located in the International Center near the airport.

Another example of Delta's commitment to Salt Lake City is its new, $19-million maintenance hangar, scheduled for completion in mid-1989, which will provide full maintenance support for Delta's extensive operations. In addition, Delta modernized and renovated its airport facilities during 1988 at a cost of about $11 million, bringing Delta's outlay for constructing and upgrading facilities to nearly $40 million—all following the merger with Western.

More than 4,000 Delta professionals, including those added since the merger to aid in staffing new facilities, live and work in the Salt Lake City area. Delta, which maintains bases for pilots and flight attendants and employs personnel in all areas of airline operations in Salt Lake City, is Utah's sixth largest employer.

Delta's direct economic impact on the Salt Lake City area exceeds $250 million annually. Approximately $60 million is paid as salaries of Delta employees, with the remainder going for landing fees, rentals, utilities, passenger food, fuel, travel agency commissions, and state and local taxes.

Wherever Delta flies, its presence helps to spur economic growth through the purchase of goods and services, through the salaries paid to its employees, and by bringing businesspeople, vacationers, and conventioneers into a city.

Delta, a major world airline with more than 2,200 flights daily to more than 150 cities, serves Salt Lake City with more than 2,300 seats daily to 72 cities.

In order to strengthen Salt Lake City operations, Delta substantially renovated and modernized its airport facilities in 1988. The project included the renovation of four gates on Concourse B obtained from America West. Areas on concourses C and D have also been updated to include a new group room and two new information counters to assist passengers, thus eliminating their having to return to the main ticket lobby.

Salt Lake City travelers also benefit from an improved flight information display system. This computerized system provides up-to-the-minute flight information with state-of-the-art technology.

Many of the improvements brought about by the $11-million program won't be visible to travelers—improvements such as computerized baggage handling and conveyor systems. And a new air conditioning system will allow planes to be cooled while they are on the ground without the use of their own power units. But travelers will notice for years to come the excellent Delta service made possible by these steps.

Delta is the third-largest airline in the free world, operating nearly 2,300 flight segments daily with a fleet of more than 390 aircraft. Carrying 60 million-plus pas-

sengers annually, Delta serves more than 150 cities worldwide. In addition, regional carriers known as The Delta Connection serve more than 120 cities in 36 states and Canada with schedules conveniently arranged to meet Delta's arrivals and departures. In Salt Lake City, SkyWest is The Delta Connection carrier, offering flights and connections to cities in Colorado, Idaho, Montana, Nevada, and Utah.

Delta, one of the nation's most profit-

Delta is proud of its customer service, and for 15 consecutive years has achieved the lowest complaint record of all major airlines, according to the U.S. Department of Transportation.

able airlines, has paid successive dividends to its stockholders every year since 1949 and has been profitable for 39 of the past 40 fiscal years.

As an airline that takes pride in its customer service, Delta has had the lowest passenger complaint record of all the major carriers for 15 consecutive years, according to Department of Transportation records.

Delta's achievements through the years have been built upon the loyal and dedicated service of its people. There is a special relationship between Delta and its personnel that is rarely found in any firm, generating a team spirit that is evident in the individual's cooperative attitude toward others, cheerful outlook toward life, and pride in a job well done. Also contributing to Delta Air Line's professionalism is the longevity of service of its personnel, resulting from the company's long-standing policy of promotion from within.

More than 4,000 Delta professionals call the Salt Lake City area home. Delta also maintains bases for pilots and flight attendants there.

QUESTAR CORPORATION

Questar Corporation's importance to the Salt Lake City area goes back to 1929, when predecessor companies brought natural gas to Salt Lake City from producing areas in southwestern Wyoming. This versatile, clean, economical fuel was welcomed by cheering throngs and is now used in nearly a half-million Utah homes and businesses.

Today Questar is a diversified energy company with 2,700 employees engaged in all aspects of the oil and gas industry. Questar common stock is traded on the New York Stock Exchange, and the corporation plays an active role in Utah economic development activities.

The best-known Questar affiliate is Mountain Fuel Supply Company, which provides retail natural gas distribution service to nearly a half-million customers. Mountain Fuel's service area includes northern, central, and southwestern Utah and southwestern Wyoming. The firm has an outstanding reputation for reliability, an excellent supply situation, and rates that are historically among the lowest in the nation.

Questar's other major lines of business are interstate natural gas transmission and oil and gas exploration and production. Questar Pipeline Company engages in natural gas sales for resale, gathering, storage, and interstate transmission activities. It operates a 2,500-mile pipeline system and gas-storage facilities in Utah, Wyoming, and Colorado.

Oil and gas exploration and production are carried out in 13 states by Questar affiliates Celsius Energy Company, Wexpro Company, and Universal Resources Corporation.

Other Questar affiliates engage in brick manufacturing, conduct business development and energy research activities, own and manage commercial real estate, and provide data-processing and microwave communication services. These affiliates are Interstate Brick Company, Questar Development Corporation, Questar Telecom Inc., Interstate Land Corporation, and Questar Service Corporation.

ABOVE: A worker checks rigging on a successful development well drilled by Celsius Energy in southwestern Colorado. Questar affiliates Celsius Energy, Wexpro Company, and Universal Resources Corporation conduct oil and gas exploration and development activities in the Rocky Mountains and the Midcontinent region.

ABOVE: Mountain Fuel is proud of its reputation for service and reliability. Since natural gas service began in Salt Lake City in 1929, the company has never had a major disruption in service in its Utah service area.

RIGHT: Five 2,600-horsepower compressors inject natural gas into one of the West's most advanced gas-storage facilities, operated by Questar Pipeline Company at Clay Basin in the northeastern corner of Utah. Questar Pipeline also engages in natural gas sales for resale, gathering, and interstate transmission.

Questar affiliate Mountain Fuel provides natural gas service to nearly a half-million customers, including recreational facilities such as Snowbird in Little Cottonwood Canyon, east of Salt Lake City. Snowbird uses natural gas-fired cogeneration equipment to generate electricity, then uses the waste heat from the generators for space heating.

INTERMOUNTAIN POWER AGENCY

In 1983 the Intermountain Power Agency (IPA), located in Murray, Utah, sold the largest single issue of municipal bonds that Wall Street had ever seen. That was the first of many records set by the agency in the financing of the $5.5-billion Intermountain Power Project. Ultimately IPA became the nation's largest issuer of municipal bonds.

The Intermountain Power Agency was formed in 1977 to own, finance, construct, and operate the Intermountain Power Project, a joint effort of 36 utilities that had all forecast additional power needs by the 1980s. Six of the utilities were located in Southern California. The Utah participants included 23 municipal utilities, six rural electric cooperatives, and one investor-owned utility. Searching for an appropriate way to finance the construction, they went to the Utah legislature for help.

The legislature responded through the Interlocal Cooperation Act, which allowed the creation of joint-action agencies for financing such projects as sewer and water systems, amending it to include power projects. The 23 Utah municipal util-

ities formed the agency, which was made a political subdivision of the state by the legislature. The legislation allowed the agency to finance the Project through the issuance of tax-exempt municipal bonds, the preferred, lower-cost financing vehicle.

The bonds would be secured by Power Sales Contracts with the participants. These and other documents, including an agreement with the Los Angeles Department of Water and Power to manage the construction and operation of the Project, were all prepared in the late 1970s.

The feasibility studies were done and the federal environmental impact statement completed before IPA was formally created. The Project at that time was to be located in central Utah.

Concerns over federal environmental policy caused the first site to be abandoned, and a search for a new site began. Utah's then-governor, Scott Matheson, formed a sitting task force, and a location 125 miles southwest of Salt Lake City, near Delta, Utah, was selected. Construction was scheduled to

begin in the fall of 1981.

The construction was divided into packages, and bids were invited. A Site Stabilization Agreement was reached with the major labor unions that allowed both union and nonunion contractors to bid competitively. The agreement included training for the local work force; ultimately, more than 70 percent of the workers were hired from the local area. Throughout the construction period more than 14 million man-hours were worked without one hour of time lost due to labor/management problems.

Money for the construction effort was already accruing in the construction fund. The agency had formed a financial team and issued its first bonds in February 1981. A second bond issue was sold in May, and by the end of 1981 the total in the construction fund reached $1.5 billion. The IPA financing plan was to enter the market in an orderly way, with issues of a size that the market could absorb to keep borrowing costs as low as possible. Construction would be prefunded to avoid costly delays in schedule if market conditions prohibited issuing bonds. The proceeds from each bond sale would be invested in high-quality government securities until they were needed.

The construction efforts that first year primarily concerned preparation of the 4,614-acre site. Excavating and filling proceeded for the railroad, roadways, waste-disposal areas, and on the areas around and under the generating buildings. Work on the substructures was well under way by the end of 1982.

In the meantime, the financing efforts had been discontinued. The nationwide economic downturn had prompted a reevaluation of energy needs, which resulted in reduced load forecasts for some of the participating utilities. As a result, negotiations began to reduce the Project by one-half, to two 750-megawatt generating units. Those negotiations lasted almost a

The 1989 Intermountain Power Agency board of directors (clockwise, from left): Clifford C. Michaelis; W. Berry Hutchings; Newel G. Daines, Jr.; John A. Mohlman; Ray Farrell; R. Leon Bowler; and Reece D. Nielsen, chairman.

year and prevented the agency from issuing bonds during 1982.

With all of the participants in agreement, the Project was officially reduced to two units in early 1983, and the agency was able to go to market with a $900-million issue. The cost estimate for the new two-unit project was $5.5 billion.

Over the next three years every construction milestone was met, and the construction work force grew to a peak of 4,500 in 1985. To reduce the negative impact of the work force on the surrounding community, the agency initiated an impact alleviation program well in advance of construction. Temporary housing was built and money was provided to local government for schools, sewer and water improvements, public safety, parks, hospitals, and operating and maintenance budgets. The "boomtown" effect common to large construction projects was prevented.

Financing continued through 1983, and by the end of the following year the initial financing was completed. The Southern California participants contributed $1.1 billion for the Southern Transmission System, and IPA supplied $4.4 billion. The overall borrowing cost of 11.3 percent reflected the inflationary economy at that time. An active bond refunding program was initiated in 1985 that lowered the borrowing cost to its present rate of 8.41 percent. Refunding involves replacing more expensive, higher interest bonds with lower interest, less expensive bonds. The refunding program resulted in debt service savings to the participants exceeding $2 billion over the life of the Project.

Another $1.5 billion was earned through the investment of unspent bond proceeds. A well-managed construction effort maintained those savings, and the Project was completed with a $300-million surplus.

The Project was officially completed on May 1, 1987, and dedicated on June 13, 1987. Ahead of schedule and under budget, it was one of the most successful construction projects of the decade.

The operating phase of the Project's history has been equally successful. The permanent work force, 90 percent from Utah, was fully trained and ready to accept responsibility when the construction was finished. The measurements of successful operation, availability, and capacity are extraordinarily high, and the cost of energy is now less than half of what was projected in 1983.

The Intermountain Power Agency will continue to manage its outstanding debts and its investment portfolio to further reduce the borrowing cost and the cost to consumers of Project power. The Intermountain Power Project's record of success continues.

UTAH POWER & LIGHT COMPANY

"The power company just sells electricity. Why would it get involved in bringing new companies and industries to the area it serves?"

This question is not unusual. Many people are unaccustomed to seeing an electric utility working to help other businesses grow. But it is happening. It is happening because it makes good sense. And more of it will be happening in the future.

It was not long ago that Utah Power & Light Company had all of the growth it could handle. Customers were demanding more and more energy every year. But the much-discussed energy crisis of the 1970s forced the situation to change. The nation's economy began to change. People began to use energy differently. These days growth does not always materialize for electric utilities. Utilities such as UP&L have to take a more active role if they want to prosper and meet the challenge of swelling ranks of competitors.

Businesses and industries consume most of the energy that UP&L generates. And whenever new businesses open, the people working there will also need power for their homes, schools, and churches. So whenever UP&L can encourage other businesses to open or expand, it has a positive effect on itself and the entire community. Central to Utah Power & Light's expanding economic development program is the idea that what is good for the community is good for UP&L.

Utah Power has created a department that works in several ways to assist community leaders, business managers, and government officials as they all strive to improve the economy.

Examples of UP&L's economic development efforts include:

Community development assistance—UP&L has been working with communities throughout its service area for several years. This is the backbone of the company's economic development efforts. Communities are taken step by step through the strategic planning and goal-setting process to establish their own direction for economic development. More than 30 communities have participated in this planning process, with some remarkable results in the past sev-

eral years.

A technical resource pool—UP&L employs highly trained engineers, accountants, managers, finance experts, and others as part of its normal business. The firm can lend those experts to businesses that may need a little temporary help with projects that may help expand or improve their operations.

Expert studies—UP&L has assisted communities by underwriting feasibility studies and target industry studies to identify the best kinds of economic development prospects. Economic development efforts are directed toward the industries that are most likely to succeed within a specific area.

Trade conferences—UP&L has co-sponsored several seminars to help businesses learn how to increase their trading ties to other nations. Conferences have focused on how to trade with Pacific Rim countries and how to take advantage of European trade opportunities.

Direct participation—Sometimes UP&L participates directly in persuading businesses to move to the area. Dr. Val Finlayson, UP&L's Salt Lake Region manager, has taken on a yearlong, full-time special assignment to help Utah attract more aircraft-related industry. He heads a team of representatives from the business and public sectors.

Utah Power & Light Company plans to continue its involvement in the economic development of communities for many years to come; it does much more than "just sell electricity."

ABOVE: A recent addition to Salt Lake City's skyline is the Eagle Gate Plaza. Encouraging new development in its service area, Utah Power & Light Company serves more than a half-million customers.

BELOW: An innovative and economical method of disbursing water used in power plants is utilized by UP&L. The company uses the water to grow crops on farms adjacent to power plants.

DESERET NEWS

Established in the summer of 1850, just three years after Brigham Young led the first Mormons into the valley, the *Deseret News'* first editions were printed using a handpress that produced two copies per minute. Typical articles included news from around the world, as well as gospel discourses and letters from far-flung missionaries. Initially published as a weekly, the publication's rates were five dollars per year, which subscribers usually paid in livestock or drygoods rather than with precious cash.

Shortly after its intrepid start the paper nearly foundered. Wagon freight from the East was painfully slow, and a lack of paper forced the *Deseret News* to publish sporadically—once suspending publication altogether for three months. But the resourceful pioneers found a way to make their own newsprint. When the newspaper could manage to put out an edition, it advertised for rags; contributions of fabric, old tents, ropes, and even wallpaper were made. A paper-making factory was then set up on Temple Square, where the raw materials were hand washed, ground between stones, and then boiled in large vats and flattened by rollers. The finished product was reportedly such a dark gray that it almost concealed the printed type. But it was paper, and the *Deseret News* marched forward.

Today the newspaper boasts a staff of top-flight reporters and photographers who use the latest technology to produce their stories. An electronic pagination computer arranges articles, photographs, and advertisements into finished pages, and photo composition equipment sets 3,000 words per minute in cold type. A $7.5-million Goss Metro offset press produces 70,000 newspapers per hour. In 1983 the *Deseret News* began publishing a morning Saturday and Sunday issue in addition to its weekday-evening editions.

The continued success of the *News* is credited to its staff. President and publisher Wm. James Mortimer is described as the "driving spirit" in the modernization of the newspaper. Renowned local writers include award-winning sports columnists, artistic photographers, political pundits, fashion columnists, and colorful artists.

The *Deseret News* invests in future reporters with an internship program and journalism scholarships offered at Utah's four major journalism schools. The newspaper also invests in future subscribers with its Newspaper in Education program in the public schools, which provides newspapers for children in the classroom to study current events.

For nearly 140 years the Church of Jesus Christ of Latter-day Saints has owned and operated the *Deseret News,* exemplifying the valley's hardy, resourceful pioneer stock, as well as its high standards and ideals.

LEFT: The staff of Deseret News *uses the latest information-gathering techniques to write and edit the news.*

BELOW: Award-winning Deseret News *reporters cover breaking stories, as well as all the news in the Intermountain West. Photo by Ravell Call*

BONNEVILLE INTERNATIONAL CORPORATION
BROADCAST HOUSE • KSL-AM • KSL-TV • VIDEO WEST • BONNEVILLE MEDIA COMMUNICATIONS

Broadcast House is home to one of the finest and most technologically advanced communication centers in the world. Located in the heart of downtown Salt Lake City, the 130,000-square-foot, ultramodern facility is headquarters for KSL-Radio, KSL-TV, Video West, and parent organization, Bonneville International Corporation. Other subsidiaries—Bonneville Media Communications and Bonneville Telecommunications—are located nearby.

Broadcast House has been carefully designed to make full use of today's technology and to anticipate the dramatic changes that will shape the future of broadcasting. It is unique in the West for the "electronic family" of companies it houses, and for the reputation and success of each of the individual family members.

KSL Radio is one of America's leading news and information stations, offering listeners a complete news, sports, information, talk, and public affairs programming. It is one of only 11 stations permitted by the FCC to broadcast via 50,000 watts of clear-channel AM stereo sound. Its powerful signal covers more than 100 counties in 10 states, and it has a nighttime skywave coverage beyond measure. State-of-the-art electronics are contained within the facility's seven studios, two voice booths, and six editing bays.

State-of-the-art electronics are contained within KSL Radio's seven studios, two voice booths, and six news editor bays.

Broadcast House's bronze-glass tower brings a modern communications center to downtown Salt Lake City.

KSL Radio is known as The News Authority, having earned a reputation as a source for complete, accurate, and timely news. The station's news staff is the largest radio news staff in the area. KSL's custom-designed news-gathering equipment allows reporters to send live, studio-quality reports from throughout the West, and combines with the unmatched news capabilities of CBS Radio to keep Utah citizens well informed about events of the world, nation, state, and community.

KSL Television is the CBS affiliate in Utah. A system of more than 100 translators carries the KSL-TV signal throughout the Intermountain West, from Montana to Arizona and from Nevada to Wyoming. Its facility at Broadcast House includes three TV studios and a TV gallery with theater seating. Only the most modern electronic systems are used in KSL's sophisticated master control room, editing bays, and videotape center.

KSL-TV's newsroom is a showplace for state-of-the-art technology. News-gathering resources include an award-winning staff of reporters and photographers, a portable satellite uplink truck, a news helicopter, mobile video recording and transmitting equipment, satellite dishes, computer-stored still photos, and the latest cameras and editing equipment. KSL-TV News has the capability of reporting live from anywhere in the world.

KSL-TV News received the Edward R. Murrow Award from the Radio-Television News Directors' Association as the outstanding news operation in North America. In his book, *Untended Gates,* author Norman Isaacs cites KSL-TV News as one of four four-star television news organizations in America. Columbia University professor Fred Friendly frequently names KSL as one of the outstanding news stations in the country. Its staff has been honored by virtually every national professional journalism organization.

In addition to news, local public affairs programs, and the CBS network, KSL-TV also provides viewers with supple-

ABOVE: *Broadcast House is KSL's and Bonneville International Corporation's commitment to the future.*

BELOW: *KSL-TV's newsroom rivals that of any information-gathering organization in the country.*

mental services. The station was the first in the nation to broadcast Teletext, which now includes a full local information service as well as CBS Extravision. Teletext, known as the newspaper of the air, allows subscribers 24-hour access to "pages" of information via their television sets. In addition, KSL-TV's stereo system enables the broadcast of Spanish-language audio on the secondary audio channel.

Both KSL Radio and Television sponsor or co-sponsor several major fundraisers each year. Events such as the Primary Children's Medical Center Radiothon raise thousands of dollars for nonprofit organizations.

Video West is the production arm of KSL. It is recognized throughout the West as an outstanding commercial production company that also develops and produces television programming, video cassettes, video training materials, and other products.

Bonneville Media Communications is a separate division of Bonneville International. Located near Broadcast House, the company is a full-service advertising specialist with an extensive audiotape duplicating facility. Bonneville Media productions have won every major award, including the first Clio Award given for a public-service announcement, and its products are seen on television screens nationwide.

Bonneville International Corporation is the parent company of KSL and stands as one of the most influential forces in the mass communications industry. The corporation also owns broadcast properties in Seattle, San Francisco, Los Angeles, Dallas, Kansas City, Chicago, and New York. It operates a music programming service in Chicago and a data-transmission company in Salt Lake.

From its beginnings as a 250-watt radio station atop the Deseret News building in 1922, KSL and Bonneville International have grown into a role of leadership in the community and the broadcasting industry. Through quality programming and the development of new technologies, its foresight has shaped and enriched the course of broadcasting and has touched the lives of millions. The Broadcast House family continues its commitment to the community and to the future of broadcasting by striving to improve the quality of the electronic media and by developing new forms of communication. The goal at Broadcast House is to help the viewing public better understand a world in constant transition.

KUTR-AM RADIO

The Salt Lake area is famous in the world of radio for its quantity and its quality.

Salt Lake is the 40th-largest media market in the country, but it has the same depth and breadth of radio as markets and cities many times its size. A quick flip of the dial tunes listeners to country/western, hard-rock, classical, easy-listening, talk and call-in, and golden oldies stations. In addition, the Salt Lake area boasts one more format, perhaps unique in all of radio. Station KUTR 860 AM Stereo programs music and features that parallel the Mormon life-style.

KUTR specifically targets the Church of Jesus Christ of Latter-day Saints popula-tion in the Salt Lake region. It is an ef-fort to harmonize with the church's beliefs, rather than promote religious doc-trine. The station's major emphasis is on music, but it also donates air time to appro-priate nonprofit organizations for promo-tion of upcoming events. KUTR believes it is important that these organizations have an outlet to reach the community.

KUTR offers a blend of traditional and contemporary music that appeals to members of the Latter-day Saints Church, although as a life-style station

KUTR afternoon announcer Wayne Carlson serves the needs of Salt Lake Valley's Mor-mon audience.

and not just a religious station. People who share the same values and interests as Latter-day Saints Church members en-joy the station as well.

KUTR first joined the airwaves on July 1, 1985, fulfilling a dream for many musicians in the community. "Latter-day Saints music" is a unique facet of church music, differing in tone and substance from gospel and Christian music. If it had to fit another classification, it would most easily group with easy-listening music. Its lyrics often promote high moral values and a love of God. In the more than four years KUTR has been on the air, Latter-day Saints music has evolved into a prosperous industry. Mor-mon artists are a small, supportive

KUTR Radio often takes crew and equipment "to the street" to broadcast live remote programs at community events.

group, but their search for excellence provides healthy competition for the best lyrics, music, and production quality. KUTR strives to play only the best of the Latter-day Saints music available. Such selections played on KUTR include the music of Afterglow, Lex De Azevedo, Janeen Brady, and Janice Kapp Perry, among others.

Monday through Saturday KUTR offers a 50-50 mix of contemporary Latter-day Saints music and secular music that has been screened for lyric content. Secular music artists include Kenny Rogers, Lionel Ritchie, Anne Murray, and Barry Manilow. All artists performing on KUTR play music that reflects all the positive aspects of the Mormon life-style.

Commercials played on KUTR are also screened for content. Products that will not be advertised on KUTR include tobacco, alcohol, casinos, or any other product or service that may contradict the Mormon life-style.

On Sundays KUTR programs a total-Latter-day Saints music selection that reflects the tone of the day. Listeners hear a more traditional sound, including music from the Mormon Tabernacle Choir, The Mormon Youth Symphony and Chorus, and hymns performed both instrumentally and vocally.

KUTR carries NBC news in the mornings and afternoons to keep listeners informed of world events. Along with national news, local news of interest to

the Mormon listener can be heard in the morning and at noon. "Mormon Journal," the public affairs program KUTR carries on Sunday mornings and evenings tackles subjects that the entire community will find of interest.

KUTR offers the Mormon population the type of musical and uplifting programming that is a natural for the state of Utah. KUTR's goal is to mirror the Latter-day Saints population and reflect, through music, all the positive aspects of the Salt Lake area's wholesome life-style.

MANUFACTURING AND MINING

P roducing goods for individuals and industry, manufacturing and mining firms provide employment for many in the Salt Lake Valley.

Mineral Mine, 140-141

Hercules Aerospace, 142-143

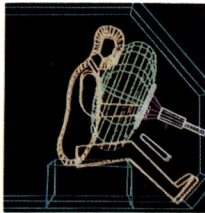

Morton International, 144

Thiokol Corporation, 145

Cytozyme, 146

Varian Eimac, 147

Geneva Steel of Utah, 148-150

Natter Manufacturing Company/Fairchild Industries, 151

Mark Steel Corporation, 152-153

Savage Industries, Inc., 154-155

Photo by Steve Greenwood

Jetway Systems, 156-157

Huntsman Chemical Corporation, 158-159

Stabro Laboratories, Inc., 160-161

Deseret Medical, Inc., Becton Dickinson, 162

Eastman Christensen,163

MINERAL MINE

Bingham Canyon Mine, the world's largest excavation, continues to play a significant role in Salt Lake Valley economy—a role it has maintained for a good portion of the twentieth century. And Kennecott Utah Copper is, with the completion of a $400-million modernization project, assured of entering the twenty-first century as a strong, efficient competitor in the world copper market.

The mammoth mine and its associated concentrator, smelter, and refinery are currently producing 200,000 tons of refined copper annually, plus substantial amounts of gold, silver, molybdenum, and other minerals. About 2,300 Utahns are employed in the production facilities

Kennecott Utah Copper's $400-million renovation project includes a state-of-the-art ore concentrator in Copperton, Utah.

and at Kennecott headquarters in downtown Salt Lake City.

To maintain this production rate, the big electric shovels in the pit have to make more than 1.4 million dips into rock blasted from the mine walls. Each shovelful—about 30 cubic yards—contains approximately 55 tons of materials. At this rate, a total of nearly 78 million tons are mined each year. Half

of the rock is ore; the rest is waste that is hauled in giant trucks to pitside dumps.

Farmer Erastus Bingham and his sons, Sanford and Thomas, gave their names to the district shortly after they obtained grazing rights in the canyon in 1848. Their prospecting unveiled "promising indications" of mineral content in rocks on the property. However, they obeyed orders of Latter-day Saints Church authorities and concentrated on food production. Other prospectors followed, and the Bingham District was formally established in 1863. That fall George B. Ogilvie, a logger working for LDS Bishop Archibald Gardner, found a particularly enticing sample. The rock was sent to U.S. Army Colonel Patrick E. Connor, commander of the Third California Infantry at Fort Douglas.

Colonel Connor had been to mines on Nevada's Comstock Lode and California's Mother Lode and recognized the potential of the sample's copper and silver content. The Jordan Silver Mining Company was formed on September 17, 1863, by Connor, Gardner, and associates. The colonel ordered soldiers to intensify prospecting in the canyon. More gold, silver, and copper deposits were located, and the Bingham and Camp Floyd Railroad was

built in 1873 to facilitate marketing the minerals.

Other mining companies were formed, and several small- to medium-size mills were erected. Among developers were such Utah mining history notables as Enos A. Wall, Samuel Newhouse, Thomas Weir, Robert C. Gemmell, Joseph R. DeLamar, Charles M. MacNeill, Spencer Penrose, R.A.F. Penrose, and Daniel C. Jackling.

Jackling, a farsighted engineer, took the lead in the incorporation of the Utah Copper Company on June 4, 1903. Water, essential to processing, was scarce in Bingham Canyon. So Jackling built a railroad around the northeastern foothills of the Oquirrhs to a new, 6,000-ton capacity mill at Magna, near the southern shores of Great Salt Lake, and to the adjacent Garfield Smelter of the American Smelting and Refining Company.

Small steam shovels began digging into a mountain at the head of Bingham Canyon—a mountain long gone and replaced by the cavernous Bingham Pit—in August 1906. It took a month to remove the first 100,000 tons of overburden, the same quantity that is now handled by Kennecott Utah Copper in a single day. The Utah Copper Company acquired the competitive Boston Consolidated Mine and its Arthur Mill in 1910, and Jackling raised capacity of the two mills to 18,000 tons daily.

The Kennecott Copper Corporation was formed on April 29, 1915, to develop Alaskan copper deposits. It was named for early explorer and naturalist Robert Kennicott, whose name was changed—the "i" was replaced by an "e"—through a recording clerk's error. Kennecott took an early interest in Utah Copper and acquired sole ownership in 1936. Eventually Kennecott was purchased in 1981 by the Standard Oil Company, which in turn was totally aquired by the British Petroleum Company in 1987.

Through the years the pit grew deeper, and the canyon town of Bingham ceased to exist on November 22, 1971. Many changes were made in methods of handling the ore and waste with the most spectacular being the $400-million modernization program, completed in 1988.

Kennecott Utah Copper boasts the biggest open-pit copper mine in the world, one of the few man-made objects discernible from space.

Giant shovels dump ore into trucks weighing 150 tons when empty; these vehicles are capable of hauling 170 tons of ore. These million-dollar trucks have six 10-foot tires and huge diesel-electric motors. They travel along roads through a 2.5-mile-wide, half-mile-deep pit covering 1,900 acres. Chicago's Sears Tower, the world's tallest building, would reach only halfway up the pit's sidewalls.

Waste goes to dumps and the mineral-bearing ore is deposited in a huge in-pit gyratory crusher that breaks the rock to a maximum 10-inch size. A six-foot-wide conveyor belt carries the crushed rock three miles through a former railroad tunnel and two more miles along the Oquirrh foothills to a 350,000-ton stockpile just north of Copperton.

The new concentrator at Copperton is a marvel of mechanical, chemical, and electronic ingenuity. A microprocessor-based control system directs movement of the ore from the stockpile on three parallel conveyors. A trio of semi-autogenous grinding mills, each 34 feet in diameter, 15 feet long, and driven by two 6,000-horsepower motors, begins the reduction process. Ball mills, 18 feet in diameter, 28 feet long, and powered by 5,500-horsepower motors, complete the grinding of the rock to the consistency of face powder.

Ninety-seven large floatation cells, including 33 each with a capacity of 3,000 cubic feet, make the first separation of copper and other valuable minerals from the ore.

Concentrates containing 28-percent copper are pumped 17 miles through a six-inch steel pipeline to the Garfield smelter where 99.6-percent pure copper anodes are formed. Copper anodes are subjected to an electrolytic process at the refinery where the copper is refined to a purity of 99.99 percent, and the gold and silver are removed. Tailings from the concentrator flow through a 48-inch concrete pipeline to a pond near Magna. Water reclaimed from the tailings pond and "makeup" water from wells, springs, and canals are pumped from Magna back up the hill to a 7.5-million-gallon reservoir just over the concentrator.

Visitors to Utah consider the Bingham Canyon Mine a major attraction in Salt Lake Valley. A visitors' center, reachable through Copperton on an all-paved road, is open from 8 a.m. until dusk, seven days a week, in the late spring, summer, and early fall. Recorded messages (available in English, Spanish, German, or Japanese at the push of a button) tell of the mine's rich history and describe the process.

Visitors are told that during the past century Bingham Canyon has yielded more than 12 million tons of copper, the value of which exceeds eightfold the combined treasures of the equally famous Comstock Lode, Mother Lode, and the Klondike. That is proof of the tremendous impact Kennecott Utah Copper has had on the state's economy. The recent modernization shows that Kennecott is confident that the Bingham Mine will continue to flourish as one of the world's most important sources of copper.

HERCULES AEROSPACE

Hercules has been a dynamic, growing part of Salt Lake City and Utah for the past 75 years.

Shortly after Hercules Powder Company was formed in 1913, the Bacchus Works plant was built in southwestern Salt Lake County to produce dynamite for the mining industry in Utah and the West. Hercules continued to produce dynamite at Bacchus until the late 1950s, when the company began to design and produce solid rocket motors for the Department of Defense.

Since the initial Minuteman Program, Hercules has provided solid propulsion for every Department of Defense strategic weapon system. Hercules continues to lead the industry in solid rocket propulsion with the development and production of the Delta II, Titan IV, and Pegasus motors for space applications.

New and innovative Hercules technologies have resulted in constant growth in facilities and employment. In the early 1960s the company expanded its operation to include the manufacture of filament-wound rocket motor cases at the Naval Supply Depot in Clearfield. In 1971 Hercules began production of graphite fiber at a site adjacent to the existing Bacchus Works. The latest addition to the Hercules Utah plants, Bacchus West, is the most modern, fully automated rocket motor manufacturing plant in the free world. Hercules' Utah operations include four plants that occupy 7,000 acres and employ approximately 4,000 people.

Today Hercules Aerospace provides reliable, low-cost solid propulsion for both defense and space applications. Hercules, with its joint-venture partner, is producing the first and second stages of the Trident-II, a submarine-launched missile, for the U.S. Navy, as well as the third stages of the Peacekeeper and Small ICBM deterrent systems for the Air Force. In addition, Hercules provides the laser fiber-optic ordnance for Small ICBM.

RIGHT: Bacchus West, Hercules' automated manufacturing facility for large rocket boosters, uses the most sophisticated technologies available for the production of safe, reliable boosters for both defense and commercial applications.

Hercules is using the efficiencies and quality advantages of Bacchus West to become a major producer of solid propulsion systems for U.S. space efforts. Delta II and Titan IV are the foundation of the country's unmanned space launch efforts. Both systems are produced at the Hercules Bacchus West facilities. The newest U.S. capability—Pegasus—is a low-cost, small satellite launch system that is being developed without government fund-

RIGHT: Using state-of-the-art design and analysis techniques, Hercules has developed and produced solid rocket motors for every Department of Defense strategic weapon since Minuteman.

Cup yacht, the *U.S.A.,* to create a rudder with weight savings of one-half to one-third over conventional designs through the use of graphite fiber. This new design helped the *U.S.A.* into the semifinals and firmly established the future of graphite composites in competitive sailing craft.

Hercules graphite fibers are used in most military and commercial aircraft, as well as naval and automotive products, golf clubs, tennis rackets, and fishing rods. The winning McLaren Formula One race car chassis is constructed of Hercules graphite/aluminum honeycomb design. This graphite chassis, which is 30 percent lighter, yet stiffer and stronger than the traditional aluminum chassis, is resistant to buckling and provides increased driver safety.

Hercules knows that the key to continued growth and prosperity is in developing new technologies and finding practical applications for them. New and innovative technologies developed today will be the Hercules product lines of the future.

ing. Hercules and its joint-venture partner are combining state-of-the-art technologies and cost-effective operations to create an entirely new commercial satellite launch system.

Hercules Utah enterprises are not limited to solid propulsion. The firm began production of graphite fiber in 1971 to take advantage of graphite's high strength and low weight for use in rocket motor cases. Rocket motor cases for the Trident-II, Peacekeeper, Small ICBM (Midgetman), Delta II, Titan IV, and Pegasus rockets are all fabricated with Hercules graphite fiber.

One of the most dramatic demonstrations of the capabilities and benefits of graphite fiber is the *Voyager,* the first aircraft to fly around the world nonstop without refueling. Hercules graphite fiber, with its unique strength and stiffness prop-

erties, allowed the *Voyager* to carry five times its weight in fuel.

Hercules engineers worked closely with designers of the 12-meter America's

MORTON INTERNATIONAL

For more than 135 years, the name Morton Salt, with the familiar umbrella girl package, has been one of America's best recognized brands. But invisible to most consumers are the many hundreds of other Morton International products that enrich their lives.

Although the Chicago-based corporation's Specialty Chemicals Division has no production plants in Utah, Morton products are essential to or are used by many of Utah's industries. One of the world's largest producers of flexible packaging adhesives and coatings, petroleum dyes, high-technology electronics, sealants, and biocides, Morton Specialty Chemicals protect and enhance a myriad of end products.

Morton, for example, is the free world's only producer of polysulfide polymers, which are the base of many of the sealants and adhesives used in the construction and building industries. Dyes for inks and brightening agents for paper find their way into most newspapers and magazines, as well as into ballpoint pens, marker pens, and other everyday items.

In Utah, the Salt Division of Morton International maintains a solar evaporation plant on the south shore of the Great Salt Lake just west of Salt Lake City. Salt for highway safety, agricultural, and industrial use is reclaimed from the brine of the Great Salt Lake, and a variety of other dissolved chemicals and minerals are extracted as well.

But Morton International's major impact to Utah's economy is a result of the burgeoning new Automotive Products Division located on land at Ogden Municipal Airport. Developed as an offshoot of the solid rocket propulsion industry, the passive restraint systems manufactured by this division will soon be found in a large percentage of the world's automobiles. Commonly called airbags, Morton's passive restraint systems were developed over a 20-year period and are now revolutionizing the industry.

The first inflator was developed for General Motors and resulted in the develop-

The Morton Salt Plant is located west of Salt Lake City near the south shore of the Great Salt Lake.

ment of driver and passenger inflators and a passenger restraint module. In 1982 initial production began on driver inflators for Mercedes Benz. In 1985 the new lightweight aluminum driver inflator was put into production for Mercedes. The Chrysler Corporation has also recently begun accepting inflator modules for several of its automobile models. Now firmly established as a market leader, the company also supplies inflators and modules to General Motors, Ford, and many major European and Asian automotive companies. From a sales level of $17 million in fiscal 1988,

sales hit $50 million in 1989 and are projected to exceed $200 million by 1991.

In 1982 Morton-Norwich Products merged with Thiokol Chemical Corporation and became known as Morton Thiokol, Inc. That association continued until mid-1989, when the by-then $2-billion corporation split into two independent companies, Morton International and Thiokol Corporation. Both companies remain major business influences within Utah.

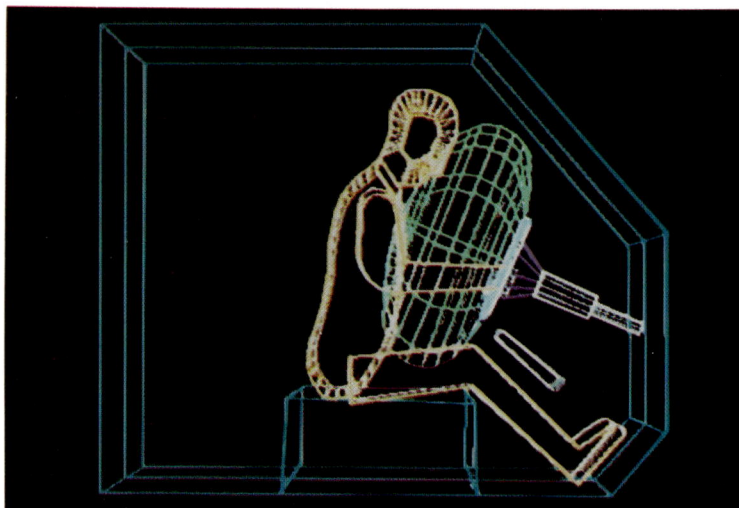

Morton International's automotive passive restraint systems are used by major auto manufacturers to protect their precious cargo.

THIOKOL CORPORATION

The space shuttle Solid Rocket Motor in this test stand at Thiokol's desert facility west of Brigham City holds more than 1.1 million pounds of solid propellant, which burn in two minutes.

In its more than three decades in Utah, the Thiokol Corporation has continued the pioneering legacy that is so uniquely the heritage and history of the state. The company came to Utah to advance the frontiers of solid rocket technology, and it has succeeded.

In 1957 the Thiokol Chemical Corporation established itself in the Utah desert some 75 miles northwest of Salt Lake City. Almost immediately Thiokol landed the contract to research, develop, and produce the first-stage motor for the Air Force's Minuteman Intercontinental Ballistic Missile (ICBM). During the 1970s Thiokol became the industry leader in solid rocket propulsion as well as a major part of the northern Utah economy; it is now the world's largest solid rocket motor manufacturing facility.

Thiokol's successful association with the nation's strategic defense continues, with production of first-stage motors for the Air Force's highly successful Peacekeeper ICBM and development of the first stage for the proposed Small ICBM, popularly dubbed Midgetman. For the Navy, Thiokol has long been a joint-venture partner with Hercules Aerospace to produce first- and second-stage motors for the submarine-launched Poseidon, Trident I, and Trident II missiles. Smaller, tactical missiles are also a major product line for Thiokol, with current contracts for the Navy/Air Force HARM air-

launched antiradar missile and the Navy's ship-launched antiaircraft Standard missile.

Other Thiokol products include complex filament-wound composite structures, military illumination flares, ordnance systems and devices, and a great variety of other solid rocket motors and products manufactured at five additional operating locations nationwide. Thiokol Corporation's national headquarters is located in Ogden's downtown business district.

But Thiokol's most well-known product is the space shuttle's Solid Rocket Motor for NASA. Since winning this contract in the early 1970s, Thiokol developed and produced the world's only reusable, segmented booster. The world's largest motor, each SRM is 12 feet in diameter,

126 feet long, and contains 1.1 million pounds of solid propellant. A pair of motors provides more than 6.5 million pounds of thrust at liftoff, almost 80 percent of the space shuttle's total thrust. These huge motors are refurbished, loaded with propellant, and either test-fired or shipped to Kennedy Space Center from the company's main plant site in Utah.

Thiokol's unprecedented effort during the complete redesign of the SRM after the *Challenger* accident resulted in the safest, most reliable solid rocket motor in the world, according to independent government agencies that monitored the redesign. Thiokol's production of redesigned space shuttle SRMs is expected to continue through the late 1990s.

With more than 19,000 acres of land, some 7,500 employees, and a sales total of more than one billion dollars annually, Thiokol Corporation is Utah's largest private industrial employer. A payroll of roughly $200 million, state tax payments of $16 million, and in-state purchases of goods and services topping $100 million each year add tremendously to the Utah economy.

Thiokol's manufacturing and test facilities encompass more than 19,000 acres of Utah's desert. The company's facilities have been continuously modernized with up-to-the-minute technologies for more than 30 years.

CYTOZYME

When one thinks of high technology, one usually thinks of microprocessors, computer chips, or the latest breakthroughs in medicine. A Salt Lake City firm has become a pioneer in another high-tech industry—agricultural biotechnology. Cytozyme Laboratories was founded for the purpose of developing environmentally safe agricultural products through biotechnological research. Today Cytozyme is a major international supplier of proven agronomic products. Its ultramodern west side plant employs 25 people who use an elaborate process to manufacture their products on a large scale.

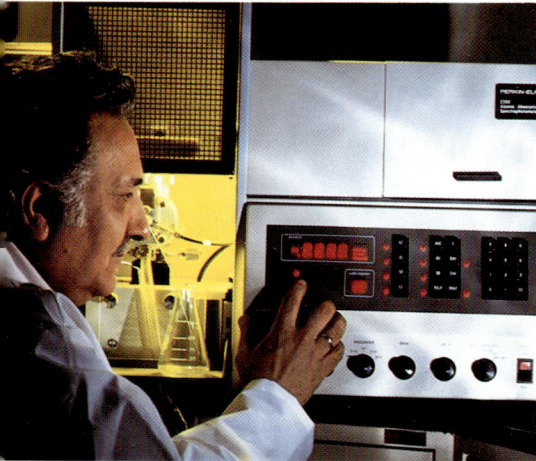

Research technicians at Cytozyme Laboratories use modern technology to create nontoxic and ecologically safe crop-enhancement products.

ABOVE: Farmers and growers worldwide use Cytozyme products to improve crop quality and produce higher yields.

RIGHT: Cytozyme Laboratories manufactures products that enrich soil by stimulating growth of beneficial microorganisms and by speeding up the process of decay and the return of nutrients to the soil.

Cytozyme founder and president Steve Baughman believes that biotechnology is the agricultural tool of the future and one of the answers to increasing the world's food supply. Almost two decades ago he began researching scientific crop enhancement, and in 1975 formed Cytozyme Laboratories to develop a line of non-toxic products. "Today more than ever, man's pursuit of happiness and a better life cannot be separated from his environment," says Baughman. "For more than 12 years Cytozyme has worked with governments, universities, research groups, and individual farmers on every continent to find better ways to grow more food of a higher quality and nutritional value."

Cytozyme's team of researchers has achieved a major breakthrough in biotechnology by developing a proprietary two-state fermentation process. In laymen's terms, what Cytozyme's array of products does is naturally promote the growth of beneficial microorganisms in the soil, seeds, plants, and animals, which produces better quality crops and farm animals.

Cytozyme products are making a difference in food production worldwide. In Asia, farmers are getting more rice, tea, and cotton per hectare; European farmers are growing more vegetables, fruit, and wheat with higher quality; and in South America crops of coffee, sugar cane, wheat, and beans have been improved. Farmers in the United States are also getting higher yields and better quality from their corn, potatoes, tomatoes, beans, and other staple foods.

Cytozyme is sold in Colombia, Costa Rica, Guatemala, Mexico, Italy, Belgium, Greece, Great Britain, Kenya, India, China, Malaysia, Indonesia, and the United States.

More colorful and fragrant flowers, such as jasmine, orchids, and chrysanthemums, thrive on Cytozyme. "Wherever the Cytozyme name appears, we hope to make that corner of the world a little more fragrant, a little more beautiful," says Baughman.

Baughman's company has come a long way and now has a significant position in the Salt Lake business community and in the expanding industry of agricultural biotechnical products.

VARIAN EIMAC

The Eimac Division of Varian, located in Salt Lake City, is the largest independent producer of X-ray tubes in the world. Planar and power grid vacuum tubes, also manufactured at the Utah plant, are utilized worldwide in such applications as radio and television transmission, industrial heating, high-voltage switches, and air navigation. Supplying high-volume customers in the United States, Europe, Israel, and particularly in Japan, Eimac has proven that with a dedicated work force, an innovative technical staff, and aggressive management it can effectively compete in the international marketplace.

Eimac was incorporated in San Bruno, California, in 1934. Two radio engineers, Bill Eitel and Jack McCullough, pooled their resources and launched Eitel-McCullough, Inc., during the depth of the Depression. Their plan was to produce high-performance electronic tubes for use in radio broadcast transmission.

World War II brought a flood of military orders, which necessitated expanded operations in San Bruno and establishment of a new facility in Salt Lake City. The Salt Lake plant quickly attained large-volume production levels and in 1943 was awarded the Army-Navy "E," symbolizing its contribution to the war effort. In 1946, following the Allied victory, the Salt Lake facility was closed. The Utah experience had made a positive impression on Eimac management, and in 1949, when product demand again justified a second manufacturing site, Salt Lake City was the immediate choice.

Eitel-McCullough was acquired by Varian Associates, Inc., in 1965, and the Eitel-McCullough facilities became the Eimac Division of Varian. Employing more than 11,000 workers worldwide, Varian is now a *Fortune* 500 company with products that include scientific instruments, medical systems, semiconductor equipment, and microwave tubes.

During the late 1970s and early 1980s the character of the Salt Lake operation changed significantly. Previously chartered for the high-quantity manufacture of power grid tubes, which had been designed and developed in California, the organization became increasingly self-sufficient with heightened engineering, marketing, and quality- and material-control capability. With its in-depth experience in vacuum tube technology, the firm believed that it could also successfully manufacture X-ray generating tubes. In 1970 the development project was launched. After an initial struggle to gain recognition in this new marketplace, the X-ray product line became a major success.

The future should provide additional challenges to the company that brought large-scale, high-technology manufacturing to Utah. Eimac's record shows that challenges can be turned into opportunities, and opportunities into successes. Those successes have been primarily attributed to the dedication and resourcefulness of Eimac's employees. The 900 men and women at the Salt Lake division are innovative people making technology work for customers worldwide. With the same confidence and commitment that pioneered the Utah electronics industry, Varian Eimac is looking forward to the challenges of the twenty-first century.

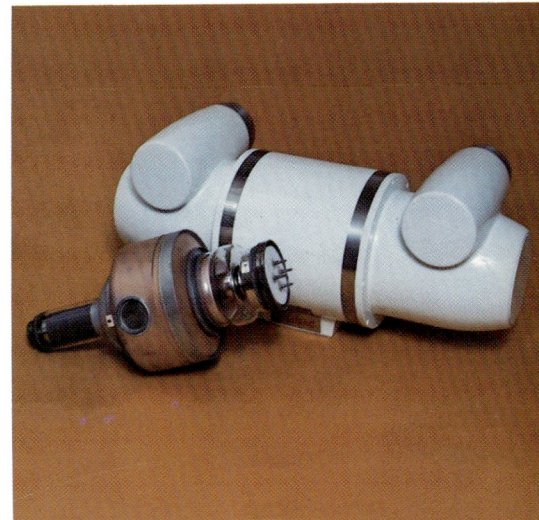

ABOVE: *Medical X-ray tubes manufactured at Varian's Salt Lake City facility are utilized worldwide.*

LEFT: *Varian power grid tubes provide the high reliability required for precision guidance systems.*

GENEVA STEEL OF UTAH

"At Geneva Steel People Make the Difference." That slogan, printed on company brochures for years, has never been more true than today.

In 1987, as a result of a depressed steel industry, Geneva Steel was idled and on the verge of closure. Today Geneva is profitable and thriving with a payroll of $55 million, responsible for more than 2,400 full-time jobs, with 8,000 peripheral jobs and more than $150 million in consumer spending in Utah County each year. The new owners credit Geneva's phoenix-like rise from the ashes to the Utah County spirit and the hard work and good faith of its people.

"Team spirit," "highly competitive," "long-term commitment"—all are phrases heard often at Geneva Steel these days. The stated goal at the plant is to prove that Geneva Steel is the most reliable and versatile supplier of quality steel in the United States and is uniquely responsive to the needs of the marketplace. "Wherever I go at Geneva Steel, I see people working together to make our plant successful," says Joe Cannon, Geneva's

president. "Innovative ideas and old-fashioned experience will result in a highly competitive facility. We need great teamwork—both inside and outside the plant—to compete successfully in an international steel market."

Owners Joe Cannon and Chris Cannon, members of a pioneer Utah family, are committed to the future of Utah County and its economic growth. Both have extensive ties to the area. Joe went to law school at Brigham Young University while his wife worked to support them as a secretary at Geneva. "The plant actually put me through school," says Cannon.

The brothers entered the steel industry through the back door. In 1987 both were in the midst of successful careers in Washington, D.C.—Joe as a corporate attorney at Pillsbury, Madison & Sutro, Chris as an attorney at the Department of the Interior. Through Utah friends and others, Joe heard that Geneva Steel was in trouble. USX, then owner of the plant, had decided to switch from Utah steel to cheaper Korean-made steel and

was considering a total shutdown of the plant. The Cannons had never aspired to be in the steel business, but their concern for Utah County and their belief in Geneva's possibilities convinced them to try to buy the plant.

Their months-long transaction makes a good suspense story—last-minute ultimatums and heart-breaking setbacks. USX had set a deadline of 5 p.m., August 31, 1987, to shut off gas at the plant. If the necessary financing could not be arranged by that time, gas to Geneva's coke ovens would be cut off. Coke ovens are a vital component of a steel plant, and their shutdown would make reopening the plant prohibitively expensive. After many months of negotiations, which included the work of scores of bankers, accountants, and attorneys, at 4:21 p.m. on the appointed day (a mere 39 minutes before the deadline) Joe Cannon handed a USX representative a check for approximately $30 million, and the deal was made.

That same night Joe and Chris Cannon, along with chief operating officer Bud Patten, performed their first duties as Geneva owners. At midnight they went out to the plant gates to welcome the first shift of steelworkers as they returned to the mill. "It was one of the high points of my life," says Joe Cannon. "I wouldn't have missed that first shift for the world."

The Cannons give credit to a team of concerned Utahns for the successful buyout. Local leaders, lawyers, accountants, professors from Brigham Young University, lawmakers (Senator Orrin Hatch was an indispensable element in the negotiations with USX), and union steelworkers who agreed to do away with costly work rules all pitched in and did what they could to save the lifeblood of Utah County. "People of diverse backgrounds came together to save Geneva from closure," relates Cannon. "They found a common interest in their emo-

Geneva Steel, near Provo, a top-10 domestic steel manufacturing company, processes raw materials (iron ore, coal, limestone, and scrap) through coke ovens, blast furnaces, and steel furnaces into finished products.

tional commitment to Utah County."

Utah's steel- and iron-making industry dates back to 1850, just three years after the Mormon pioneers first settled the region. Anxious to become self-sufficient, a group of pioneers journeyed by oxcart to an area near what is now Cedar City, where known deposits of iron and coal were located. They built a crude blast furnace, and by 1852 produced the first pig iron made west of the Mississippi River. This operation and several others that followed supplied the pioneers with enough iron to make needed stoves, grates, iron pots, frying pans, buckets, and tools.

The first successful iron ore company was begun in 1922, when Columbia Steel Corporation built a 450-ton blast furnace at Ironton, south of Provo, Utah. The first iron was produced there on April 30, 1924. USX purchased Columbia Steel in 1930, and for more than 30 years the Ironton plant enjoyed financial success. In 1963 the facility was shut down and the buildings donated to Brigham Young University for possible use as an industrial park.

In 1941 the federal government and USX combined forces for the development of Geneva Steel. The government, in the midst of war, wanted a major steel plant located in the West as insurance against loss of the vital material in the event of closing of the Panama Canal by enemy attacks. The Utah site was chosen for its inland protection from Japanese bombers, its adequate transportation facilities to the West Coast, as well as its minimum distance from iron ore, coal, limestone, dolomite, and fresh wa-

ter. USX, then known as U.S. Steel, designed and built the plant without charge. It was named after the little summer resort of Geneva on the shore of Utah Lake. The first pig iron was smelted at the new plant in January 1944, followed by the first open-hearth steel a month later.

At the end of the war, U.S. Steel purchased Geneva Steel from the government and took over operations on June 19, 1946. Geneva expanded and prospered for more than 40 years. Considered an anomaly by many in the industry —"the steel plant in the middle of nowhere," far from the shipping and manufacturing centers on either coast—Geneva survived because of the high productivity and efficiency of its workers. An early company brochure states, "At Geneva each employee has committed himself to do his job right the first time . . . resulting in im-

proved quality of production, increased customer service, and more efficient plant operation—proof that at Geneva Works, people really do make the difference."

But in the mid-1980s troubled times arrived for Geneva Steel and the entire U.S. steel industry. Subsidized foreign competition escalated, and American firms' profits dropped sharply. Many older U.S. plants were declared obsolete and closed. A work stoppage idled Geneva on August 1, 1986, and six months later USX announced plans to discontinue production in Utah.

Today Geneva is among the top 10 domestic steel-making companies in the nation. Geneva operates the only fully integrated steel mill west of the Missis-

ABOVE: Geneva's one-millionth ton of steel, produced just 10.5 months after the first shipment, lies on this train—evidence of the firm's exceptional efficiency.

TOP LEFT: Molten iron is smelted from iron ore, coke, and limestone in Geneva's three blast furnaces. It is then poured into steel furnaces and carefully refined into hundreds of types of steel.

LEFT: Molten steel—350 tons of it—is poured from a steel furnace into a ladle, to be finished into products that will ultimately be made into anything from cans to bridges.

sippi. In layman's terms, "fully integrated" means the raw materials (iron ore, coal, limestone, scrap) are processed through the various stages into semifinished and finished products. Geneva's main finished products are plates, hot rolled sheets and coils, welded steel pipe, pig iron, coke, blast furnace and open hearth slag products, coal chemicals, and ammonium sulfate fertilizer especially good for western soils.

Once again Geneva's future is bright. The company has done away with USX corporate restrictions, which severely limited Geneva's selling power. Markets have been expanded worldwide; new union agreements have been negotiated as well as lower-cost contracts with suppliers. The results are that sales in the company's first year are more than twice as much as planned. The plant shipped its one-millionth ton of steel only 10.5 months after its first shipment on October 3, 1987. Geneva is attempting to lower its future costs and expand the industrial base in Utah County by attracting steel-related industries to the area. For example, the plant has been instrumental in locating a new pipe company in the county, which will use Geneva's steel in its manufacturing process.

There is a new sense of community responsibility at Geneva. Virtually all owners, directors, and managers live in Utah County and have the same con-

Geneva manufactures line pipe from 6 to 16 inches in diameter, especially suitable for the transmission of natural gas and oil, integral components of Utah's economic development.

Plate and strip (coils and sheets of steel) products are rolled to customer specifications on Geneva's 132-inch rolling mill. The finishing stands (above), with their computer and automated controls, provide quality products.

cerns for their valley as all other residents. Geneva meets stringent air-quality standards set by the Environmental Protection Agency. "Our plant, and the steel industry in general, is confronted daily with concerns about the environment. We share those concerns," says Joe Cannon. "As members of the community, we feel a special obligation to maintain compliance with environmental standards." Geneva has $200 million worth of environmental controls at the plant and spends $40 million per year in environmental compliance. Numerous tests by the Utah State Health Department have shown the plant to be in compliance with EPA standards.

Geneva has a strong commitment to the future of Utah County. The plant has a business support partnership with education and has sponsored numerous cultural events, including concerts by the Utah Symphony, performances by Ballet West, Springville Art Museum exhibits, and the annual Freedom Festival.

"The new team's mission is to make Geneva's steel-making operation competitive and to contribute to a safe, clean community and sound economy. We feel we're in partnership with Utah, building a better future," explains Cannon. "Our future looks bright, and we have stable markets for our steel. But our major asset is our fabulous work force—the best in the world." At Geneva Steel the people really do make the difference.

NATTER MANUFACTURING COMPANY/FAIRCHILD INDUSTRIES

Natter Manufacturing Company fills an important niche in the metal-fabrication industry. The firm's products include kiosk-type telephone booths, computer frames, and circuitry cabinets—products that require durability, uniformity, flawless paint finishes, and, most important, dimensional precision.

Employing the heavy use of computers and robotics for its manufacturing process, Natter passes flat sheets of metal through its intricate array of punching, forming, and welding machines until they emerge with client-specified shapes and imprints. "Natter's products must pass extreme accuracy tests," says Jerry Koontz, vice-president/operations. "We demand tolerances measured in thousandths of an inch." The thousands of wires and circuits that operate within a computer require an exact match of circuits and housing in order to operate correctly.

After Natter's products have been punched and formed, they go through the company's intricate painting process. Specially designed 1,500-foot conveyorized lines wash, dry, and paint the products in individual booths. High-temperature bake ovens complete the cycle. "The painting process is quite detailed," says Koontz. "It takes more than two hours for completion." Natter also provides silk-screening services for clients.

Natter supplies its product to major computer, medical, telecommunications, and commercial companies, including IBM

and AT&T, and ships its products to many locations nationwide and including destinations as far away as Puerto Rico.

The facility utilizes a continuous on-line computer program called Manufacturing Resource Planning (MRP) to ensure that finished products meet clients' timetable demands. MRP tracks each Natter order through assembly, fabrication, and the painting process, and is an important component in the company's achievement of the highest standards for its clients.

Natter Manufacturing, with its sister plant in Temple City, California, is one of the largest precision sheet-metal fabricators in the United States. The firm began in 1952 in Los Angeles as a tool and die business, progressing to a stamping company by 1960. The fabrication of stainless-steel cabinetry was added in 1963. Eight years later Natter added a modern paint line and became a fabricator of painted exterior panels and cabinets for

the computer industry. The continued success in this area of business led to the start-up of the Utah facility in 1979. Natter was acquired by Fairchild Industries, a *Fortune* 500 company headquartered in Chantilly, Virginia, in 1980.

The Utah plant, located at 4080 West 8540 South in West Jordan, employs nearly 200 people. With the Temple City plant it produces revenues of $30 million annually. Les T. Hansen, president of the company, credits Natter Manufacturing Company's skilled employees with its continued success. Emanuel Fthenakis, chairman and chief executive officer of Fairchild Industries, recently congratulated West Jordan workers on achieving one million hours worked without a lost-time accident.

In-depth know-how, computerized programming, robotics, and MRP systems assure each customer cost-effective efficiency with very high quality.

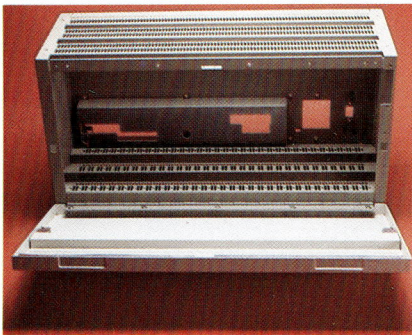

ABOVE: This Natter computer circuit chassis houses communications control circuitry. Natter Manufacturing's products are tested to meet stringent factory specifications.

RIGHT: Vice-president/operations Gerard Koontz with a communication network cabinet manufactured by Natter for AT&T.

MARK STEEL CORPORATION

Mark Steel Corporation is a custom fabricator of steel and other metals. It carries on a tradition of metalworking and electric welding that began in Salt Lake City in 1914.

That was the year that John Lang, a young German immigrant, opened Utah's first electric welding shop to serve the needs of the state's then-embryonic indus-

tries of petroleum refining, oil and gas distribution, sugar producing, chemicals, copper, coal, power plants, and construction. The Lang Company ceased operation in 1962, due mainly to a disinterested eastern owner who had purchased the firm in the late 1950s. Through the years Lang had trained many entrepreneurs who later started their own steel fabricating firms.

One of those entrepreneurs was Abraham Markosian, who founded Mark Steel Corporation in 1968. His largest plant, lo-

LEFT: Abraham Markosian, founder of Mark Steel Corporation.

BELOW: The mine modernization project at Kennecott Copper was a prodigious challenge in which Mark Steel successfully participated. The firm provided steel for an expansive conveyor system. Pictured is the discharge end of a conveyor and the upper drive station, which provides power for the system.

cated at 1230 West 200 South, was the Lang Company's principal fabricating facility.

Markosian, the son of immigrant Armenian parents, was born and raised in Murray, Utah, and is a graduate of Murray High School. While serving on a destroyer during World War II, he was selected for officer training and, under the Navy V-12 program, was educated at Tulane University and Harvard Business School, where he was commissioned an ensign. After serving his military obligations, Markosian returned to Tulane University where he received his degree in business administration in 1947. That year he began his employment with the Lang Company.

Markosian quickly moved up to assistant plant manager at Lang. In 1960 he left the firm and began a series of jobs in the steel industry, which led to his founding Mark Steel Corporation. Today, as the owner of a successful business, Markosian shares his expertise with the community through local industry and

education activities. He serves as president of the Salt Lake Community College Foundation and as chairman of the Utah Manufacturers' Association.

Now beginning its third decade of operation, Mark Steel is the largest custom fabricator in Utah. Its three plants, all located in Salt Lake City, total 24 acres and 320,000 square feet of shop and office space. The company employs a work force of 250.

Mark Steel ships an average of 1,000 tons of fabricated metal products per month. It markets aggressively throughout the United States and Canada, and has begun work for a major Japanese construction machinery company, with plans to support its sales on the North American continent.

The majority of Mark Steel's production is shipped to industrial and commercial sites in the western states, but some of the more highly engineered and labor-intensive products have been transported to the Pacific Islands and Alaska. Recently, truckload-size catalytic converters for cogeneration power plants were shipped to New Jersey and Long Island, New York. Another sophisticated steel product currently being fabricated is destined for Kentucky.

One of the most exciting and innovative of Mark Steel's recent projects is the ferry boat *John Atlantic Burr*, now based at Hall's Crossing on Lake Powell. The company's skilled work force fabricated the entire boat in Mark Steel's Jordan River plant, then trial assembled it on an outdoor lot, fitting it with tanks, plumbing, electrical, and superstructure.

After complete inspection and a dry-land commissioning by Mark Steel employees and their families and guests, the vessel was taken apart in preplanned sections and shipped more than 300 miles by truck to Lake Powell's Bull Frog Resort where final assembly, launching, and sea trials were conducted. After shakedown and acceptance by the U.S. Coast Guard in 1985, the vessel was placed into service and continues to provide the vital link in connecting Utah Highways U-276 and U-273.

The *John Atlantic Burr* measures 44 feet by 100 feet, with a carrying capacity of 2 buses, 8 cars, and 150 passengers. It was fabricated from steel produced by another group of Utah's exceptional workers, from the former U.S. Steel plant at Orem, now operating independently as Geneva Steel. The boat is an excellent example of a 100-percent made-in-Utah product.

The early and mid-1980s proved a difficult period for the country's heavy industry. The collapse of world oil prices caused a severe depression that for a time closed Utah's industrial giants, U.S. Steel and Kennecott Copper. Mark Steel managed to survive this period and even prosper, largely due to its skilled, talented, and dedicated work force, and to its ability to solicit and win production contracts from diverse economic interests. One such key contract was with the Air Force. The firm was contracted to provide Upper Launch Tube Liners for the 50 Peacekeeper sites for the base at Cheyenne, Wyoming. Boeing Aerospace Division was the commercial contractor for the project.

In 1984 a major change was made at Mark Steel's Structural Steel Fabricating Plant at 450 West 600 South. The plant was completely retooled with an automated structural steel drill line, blasting machine, and paint facilities. The changes reduced labor costs dramatically and placed Mark Steel in a position to compete anywhere in the western market.

Today Mark Steel is in top form to serve the current heavy demand for custom metal fabrication. The company has just completed participation in Kennecott Copper's mine modernization project, furnishing 7,000 tons of fabricated steel for the Grinding Building, Flotation Building, conveyors, and process tanks. Currently in process is work for Hercules Aerospace's multimillion-dollar expansion. Mark Steel is furnishing the fabricated steel for the cast cure building, case preparation building, and four solid-fuel rocket casting bell vessels. The steel required for this project totals 1,840 tons.

Mark Steel Corporation's history of adaptability and quality production has made it a leader in the current custom metal fabrication boom. The company looks forward to Utah's bright economic future, and to participating in the state's continued growth.

SAVAGE INDUSTRIES, INC.

More than 40 years ago a young Utah man and his father created a small company with a single truck, a commitment, and a dream. The truck was a KBS-5 International. The commitment was to build a sound business based upon customer service. The dream of success and expansion is still being realized. Today the Savage network of companies, wholly owned by three Savage brothers, conducts the construction and transportation industries, creating hundreds of jobs for Utahns and other westerners, and contributing millions of dollars to the local economy.

The Savage Industries dream began in 1946. Following World War II and upon completion of his service in the U.S. Navy, the oldest brother, Kenneth, returned to his family home in Utah. Determined to start his own business, he joined with his father and purchased a truck suitable for hauling heavy loads. C.A. Savage and Son, the first Savage company, officially began operations.

In those early years the firm hauled coal from Utah's rich coal fields and hay from Idaho, sawed and hauled timber from Wyoming, and hauled mine props to some of the original Utah mines. As the business grew, Kenneth's younger brothers, Neal and T. Luke, became increasingly involved in the family enterprise. They shoveled coal after school and spent their summers camping in West Yellowstone, cutting timber for customers.

By 1958 the company fleet had expanded to three trucks, and all three brothers were actively involved in virtually every facet of the business. That year the elder Savage passed away, and Savage Brothers Company was incorporated and a long-term family-owned business began.

Today, through its operating divisions, Savage is involved in ready-mix concrete, the transportation and handling of bulk commodities, specialty heavy hauling, sand and gravel, manufacturing, and property development. Savage operates the largest coal-transportation company in the United States, and is an industry leader in the transportation and handling of bulk and specialty commodities.

Savage has distinguished itself as a to-

tal transportation systems company. The Savage reputation is based upon its ability to move products through the entire transportation infrastructure, from the originating source to the consumer.

The company is one of the largest independent bulk material transportation firms in the nation, specializing in specific commodities, equipment designs, and locations.

Savage operations are managed on a dynamic, interactive basis wherein equipment and expertise are constantly interchanged to provide the most efficient service to customers. The owners

Savage trucks shown hauling coal in central Utah. Savage operates the largest coal-transportation company in the United States.

and senior management of Savage are committed to utilizing the experience gained through its decades of operations in the planning and implementation of hauling and materials-handling opportunities. The innovations and service provided by Savage during the years have helped to set the firm apart from its competitors.

The company plays a dominant role

in the construction materials industry in Utah. Through its susdiaries, Ideal Concrete and Western Rock Products, Savage is the largest ready-mix concrete supplier in Utah. Ideal Concrete and Western Rock operate more than 150 modern, front-discharge ready-mix trucks. In addition, Western Rock Products Corporation ranks among the two largest hot-mix (asphalt) firms in Utah, producing and delivering in excess of 100,000 yards of product annually.

Another subsidiary, Savage Rock Products Company, operates closely with Ideal Concrete in supplying all of the sand and gravel requirements of the ready-mix plants. In addition, it supplies sand, gravel, and road base materials to external ready-mix and road construction firms. Savage Rock Products crushes more than 2 million tons of product per year and ranks as the dominant aggregate supplier in Utah.

In January 1989 Savage sold its subsidiary, Savage Manufacturing Corporation, to Mack Trucks, Inc., of Allentown, Pennsylvania. Savage Manufacturing Corporation specialized in the design and manufacture of state-of-the-art, front-discharge ready-mix trucks and materials-handling equipment. It supplies approximately 10 percent of the front-discharge ready-mix trucks sold in the United States each year and has modernized its facilities for rapid expansion of its business in the future. Savage retained the part of the manufacturing business that has long been involved in the design and manufacture of materials-handling equipment, such as screw and belt conveyors, bins, hoppers, transloading facilities, feeders, chutes, and storage silos.

In order to handle internal construction needs and to coordinate the acquisition and marketing of company real estate, Cornelius Development Corporation was established in 1982. In addition to overseeing internal construction and de-

Cornelius Development Corporation, a subsidiary of Savage Industries, Inc., recently opened the beautiful Willowcreek Oaks community.

ABOVE: Savage's Ideal ready-mix concrete plant in Salt Lake City is part of the state's largest ready-mix chain.

RIGHT: This truck is part of Savage's fleet of pneumatic trailers—the largest fleet in the Intermountain West. Bulk commodities, such as cement, nitrate, and soda ash, are contained and applied via spray from the pressurized trailers.

velopment, Cornelius is very active in the commercial and residential real estate market.

With diverse operations throughout the United States, with modern and well-maintained equipment and facilities, and with top-flight management and more than 1,000 employees, Savage has established itself as a leader in transportation and materials handling.

The commitment that started with one Utah family is now shared by the hundreds of families whose fathers, mothers, and children work for the Savage companies. That commitment is to always look for better ways of doing things and to always follow the Savage Industries, Inc., creed: "It is our purpose to earn the respect, confidence, and loyalty of our customers by serving them in such a manner that they profit from their association with us."

155

JETWAY SYSTEMS

Jetway Systems, headquartered 35 miles north of Salt Lake City in Ogden, Utah, is the world's acknowledged leader in the design, manufacture, and installation of passenger boarding bridges for the international air transportation industry.

In 1958 the company was commissioned by United Airlines to design and build an apparatus that would allow safe, secure, and comfortable airplane passenger access, while providing for improved aircraft turnaround time. The first successful passenger boarding bridges were installed by Jetway Systems for United Airlines at New York, San Francisco, and Chicago. The first bridge, built 30 years ago, is still in daily operation in New York City (LGA).

The bridges were trademarked JET-WAY®, a name that rapidly became known for quality, dependability, and innovation—a tradition that continues to the present day. Currently there are more than 3,000 Jetway Systems' bridges at 150-plus airports in more than 25 countries worldwide. More air carriers, airport authorities, and governments worldwide turn to Jetway Systems for its products, airport planning, product installation, maintenance, training, and support services than all other manufacturers combined.

Best known for its passenger boarding bridges, Jetway's products and services today represent "more than a bridge." In the rapidly changing air transport environment, the company is known as an innovator in providing other products and services, many of which can be integrated with the bridge.

The JETPOWER® 400-hertz ground power unit was the first solid-state ground power unit introduced to the commercial aircraft industry. This small lightweight unit was specially designed to fit underneath the bridge near the aircraft, minimizing congestion and clutter on the ramp. It converts normal 50/60-hertz electrical power to the 400-hertz aircraft power that runs all on-board electrical systems while the aircraft is parked at the gate or in the hangar for maintenance.

Today's JETPOWER® ground power unit is the most efficient 400-hertz power inverter available, an important factor for airport authorities and airline operators.

Jetway's product mix has extended to other areas that also use the bridge as a link for service to the aircraft. In addi-

The first successful passenger boarding bridge was manufactured for United Airlines by Jetway Systems. Shown here after its installation in 1959, the bridge is still in daily operation.

tion to ground power, there is potable water, preconditioned air, and other equipment that helps airline and airport operators more efficiently handle aircraft on the ground, address security requirements, and reduce apron clutter.

JETLINE® aircraft ground support equipment, a full line of utility tractors, aircraft push-back tractors, and cargo/baggage tractors; mobile belt loaders; and trailers, are internationally respected for excellent design and reliable service. This equipment is designed to withstand extreme environments and rugged use in all parts of the world.

The company also designs, manufactures, and installs JETCELL® Ground Run-up Enclosures. These specially designed sound-suppression facilities help municipal airports and carriers solve sensitive problems associated with aircraft run-up noise. Jetway Systems has designed and built sound-suppression facilities for the military's jet fighter aircraft for more than 15 years. Enclosures built for commercial aviation enables air

There are JETWAY® passenger boarding bridges at more airports worldwide than those of all other manufacturers combined. Jetway Systems sets the industry standard.

carriers to avoid many political problems as well as the high costs of curfews, flight delays, and inefficient engine run-up techniques currently used at many maintenance locations worldwide.

Jetway Systems also provides many services to airport operators and airlines. Its airport planning and consultation service provides access to its extensive experience in airport planning and apron design. The firm markets custom-designed computer systems to monitor bridge and other equipment use.

Jetway Systems performs preventive and regular maintenance and service on virtually all airport facilities and equipment at many airports nationwide. The company has saved its maintenance clients significant sums on a wide variety of

projects with a management technique it uses that is unique to most airports. It currently performs maintenance on baggage claim systems, belt loader and conveyor systems, 400-hertz motor generator central systems, passenger boarding bridges, tow-tractor and ramp vehicles, preconditioned air and general terminal building maintenance including HVAC systems, as well as electrical and plumbing systems.

Jetway Systems has grown and thrived in Utah. The special blend of community values, a good work ethic, and a highly educated work force has been the company's secret in a highly competitive industry. Utah's environment and the opportunities it affords have made a great home for Jetway Systems' executives, employees, and their families.

TOP: Powerful JETLINE® Series aircraft tow tractors are built to handle heavy aircraft. Designed to perform numerous airport tasks, Jetway vehicles come in many shapes and sizes.

ABOVE: Jetway's maintenance division employs hundreds of workers who repair and maintain virtually all physical facilities and equipment in and around the airport.

LEFT: This JETCELL® ground run-up enclosure, located at Palm Springs Regional Airport in Southern California, was designed and built by Jetway Systems. The facility has reduced nighttime noise complaints by more than 98 percent and has increased maintenance efficiency.

HUNTSMAN CHEMICAL CORPORATION

Huntsman Chemical Corporation and its affiliated businesses are engaged in the production of petrochemical products, with 1988 revenues in excess of one billion dollars. These businesses produce products that are used to make thousands of consumer items that touch virtually every part of daily life.

The history of the Huntsman businesses dates back to the mid-1960s, when Jon M. Huntsman became associated with, and soon thereafter president of, the first polystyrene plastic packaging company in America. Driven by the entrepreneurial spirit that built this great country, he left this position in 1970 to form Huntsman Container Corporation.

As a research-intensive, market-driven company, Huntsman Container invented the hamburger "clam shell" and pioneered more than 80 innovative packaging concepts that are still in use today throughout the world. As the firm grew it built state-of-the-art production facilities throughout North America, Europe, and Australia.

As one of the largest U.S. consumers of polystyrene (the raw material used by Huntsman Container Corporation) during the 1970s, Jon Huntsman saw a unique opportunity for an entrepreneur to compete with the oil companies and major chemical conglomerates then controlling the polystyrene industry. He saw the need to provide more consistent product quality and service, which the U.S. polystyrene industry needed to compete in the more competitive global marketplace. This idea was accomplished through the formation of Huntsman Chemical Corporation in 1982, established to purchase the polystyrene division of the Shell Oil Company. Today Huntsman Chemical Corporation is the largest polystyrene producer in North America, leading an industry that includes such noted competitors as Dow Chemical, Mobil, Amoco, and Chevron.

Through further acquisitions and joint ventures, Huntsman businesses have expanded their role in the petrochemical industry. They now also produce styrene monomer, expandable polystyrene, polypropylene, and specialty compounded chemical products.

Styrene monomer is a liquid petrochem-

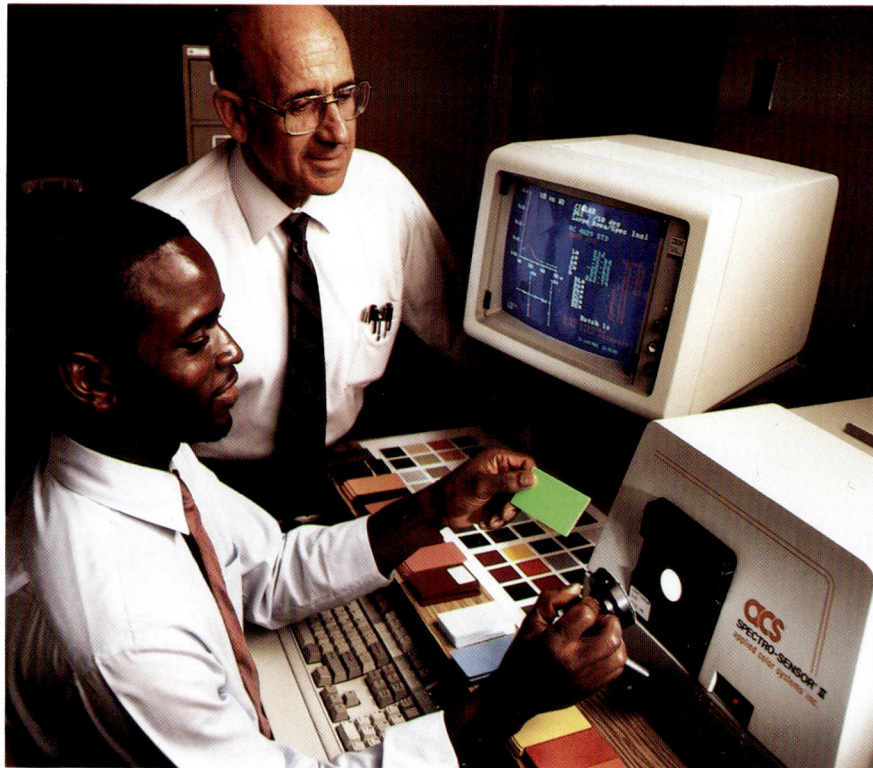

Computer-aided research and quality control are hallmarks of Huntsman Chemical Corporation.

ical utilized in the production of polystyrene and expandable polystyrene. Polystyrene is used to make a variety of products ranging from RCA television sets to egg cartons, from Polaroid cartridges to medical disposables, from Fisher-Price Toys to the familiar "clam shell" used by McDonald's. Expandable polystyrene is converted into insulation for residential, commercial, and industrial construction; protective packaging; and hot/cold cups. Polypropylene is a raw material used to make products ranging from disposable diapers to car batteries, from bottle caps to carpet fibers. Specialty compounded products have very unique qualities that meet specific needs of the automotive, electronics, and appliance markets.

Huntsman Chemical and its affiliated businesses now have 16 production facilities located throughout the United States, Canada, the United Kingdom, Taiwan, and Thailand. Further facilities are being developed in such locations as the People's Republic of China, India, Iraq, and the Soviet Union.

Huntsman Chemical Corporation reflects the character of its founder, chairman, and chief executive officer, Jon M. Huntsman, who is committed to the princi-

ple of "making things better." This is evident in his private life as well as in his business endeavors.

In 1970, after serving as a gunnery officer in the Navy, he was appointed associate administrator of the Social and Rehabilitation Service of the Department of Health, Education, and Welfare. In 1971-1972 he served as special assistant to the president of the United States.

He has since been active in the Republican party, serving as state, regional, and national Finance Committee chairman, Utah National Committeeman, and Utah chairman or co-chairman for the 1980, 1984, and 1988 Reagan or Bush campaigns.

His civic contributions include his present chairmanship of the University of Utah's $150-million endowment campaign, membership on the Board of Overseers of the Wharton School of Finance, and trusteeship of the University of Pennsylvania, his alma mater. He is currently vice-chairman of the United States Chamber of Commerce. He recently completed a three-year term as chairman of the Utah Symphony board and serves in various other civic and community causes.

The Huntsman family has been a major financial benefactor of the Utah Symphony, Ballet West, Utah Opera Company, the University of Utah, the University of Pennsylvania, Brigham Young University, the Salt Lake Homeless Shelter, Primary Children's Medical Center, and many other medical and humanitarian causes, including assistance to victims of floods in Thailand and the Armenian earthquake.

A devout member of the Church of Jesus Christ of Latter-day Saints, Jon Huntsman has served as Washington, D.C., mission president, stake president, and in numerous other church positions.

Notwithstanding his extensive church, civic, educational, and humanitarian commitments, Jon M. Huntsman is a devoted family man who consistently carves out prime time to spend with his wife and nine children.

STABRO LABORATORIES, INC.

Hundreds of companies try to measure up to the standards of Stabro Laboratories, Inc.

Stabro is in the business of calibration, the science of metrology. The company's ultramodern, temperature- and humidity-controlled laboratory houses hundreds of calibration devices. Stabro's cadre of highly trained calibration technicians utilize this equipment to certify the accuracy of myriad types of equipment— practically anything that makes a measurement and needs to hold a degree of accuracy.

"Stabro is an independent Precision Measuring Equipment Laboratory (PMEL)," explains Clark Reber, owner of the company. "We repair and calibrate our clients' electronic and mechanical test equipment. In very simple terms, we might check a thermometer company's readings to make sure their product's 72 degrees is the same as everyone else's 72 degrees."

"High-tech companies, particularly in the electronics and aerospace fields, need exact standards of measurement," says David Ebbert, general manager of Stabro. "They must continually calibrate their test equipment to ensure they are receiving accurate readings."

In addition to electronic test equipment, Stabro calibrates and maintains all types of mechanical measuring instruments, such as gauge blocks, dial indicators, micrometers, torque wrenches, and granite flats, as well as pressure/vacuum gauges, temperature devices, and medical instrumentation.

Stabro regularly certifies the accuracy of its own equipment with the National Institute of Standards and Technology (formerly the National Bureau of Standards) in Washington, D.C. Stabro is then able to provide a direct, traceable line from NIST to the customer. "NIST sets the standards for everyone," relates Ebbert. "It is the bottom line for all calibration." Stabro's equipment is calibrated by NIST, so in turn it can guarantee its clients absolutely accurate readings.

"Well not absolutely accurate," laughs Reber. "The digital voltmeter has an accuracy of .00001, or 10 parts per million. But that seems to be enough to keep the world running."

The science of calibration has come a long way. "With the development of the marketplace, thousands of years ago, people found a need for uniform measures," says Ebbert. "In the beginning the standard for length was the distance from the tip of the Pharaoh's nose to the tip of his outstretched finger. In those days merchants must have prayed for a pharaoh with short arms, and consumers must have hoped for the opposite."

Measurement, of course, improved dramatically over the course of history. However, by today's standards it remained fairly haphazard until the development of the railroad system roughly 150 years ago. "With the railroads came a need for exact timing, which necessitated the time zones," explains Ebbert. "It also brought a demand for uniform weights and measures. In the past 40 years rapid evolvement in the science of metrology has changed the approach and attitude of calibration in many facets of our lives."

One of Stabro's biggest customers is Litton Guidance and Control Systems, which utilizes eight Stabro employees in house on a full-time basis. "Our satellite operation at Litton includes a shuttle that runs equipment between our lab and Litton at least twice a day," says Reber. Stabro also maintains satellite operations at Unisys, Iomega, and National Semiconductor. Other customers include Rockwell, Symbion, Evans and Sutherland, Eaton-Kenway, the Federal Aviation Administration, Argonne National Labs, Utah Power & Light Company, and the Intermountain Power Project. The company has four vans that can test equipment on site just about anywhere. "Our mobile laboratories can perform calibration tests at mountaintop transmitters," notes Reber.

Stabro has more than $1.5 million worth of test equipment, an extensive

ABOVE: Stabro Laboratories' headquarters is located at 25 Kensington Avenue in Salt Lake City.

LEFT: An electronic calibration specialist certifies a state-of-the-art digitizing oscilloscope.

FACING PAGE, TOP: From left: owner Clark Reber, assistant general manager and office manager Sheryl Lewis, and general manager David Ebbert.

FACING PAGE, BOTTOM: A mechanical/ dimensional specialist makes a precise measurement using an optical comaritor, just one of many specialty calibration devices at Stabro.

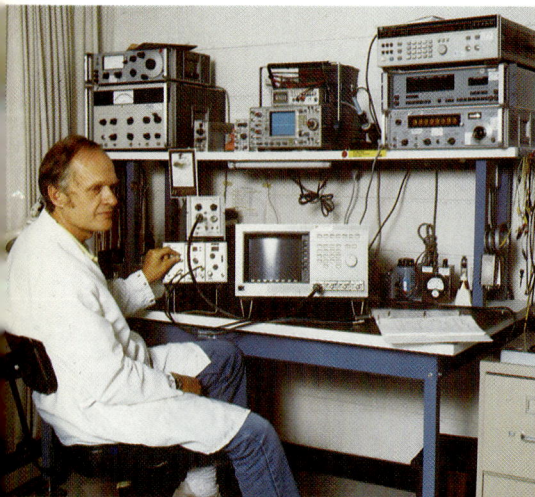

parts inventory, and a growing library of service manuals currently in excess of 7,000. The firm also maintains laboratories in Grants Pass, Oregon, and Chicago, Illinois. The company's goal is to find the problem, make repairs, perform calibration, and get the equipment back in use as quickly as possible.

A Stabro subsidiary, Intermountain Instruments, Inc., rents and sells new and like-new electronic test equipment, as well as repairs video cameras and recording equipment.

Stabro is named for the firm's founders, Jim Stahnke and Jay Brown, who formed the business in 1959. Clark Reber and Udell Campbell purchased an interest in the company in 1970, and Reber acquired it wholly in 1986. Raised on a small farm in Mesquite, Nevada, he attended the University of Nevada at Reno and graduated with a degree in agriculture education. He received a commission in the U.S. Army through finishing the Reserve Officer Training Corps program and served seven years on active duty, including a tour as a helicopter pilot in Vietnam. Reber left active duty in 1967 and is now a lieutenant colonel in the Army Reserve. He also owns an insurance business, and has

served two terms in the Utah House of Representatives.

Reber refers to his general manager, David Ebbert, as "a pioneer in metrology, who worked his way up through the ranks." Ebbert served in the Air Force for 20 years. In 1952 he trained as a radio operator, and was one of the first Air Force trainees to take equipment calibration courses. He worked in Air Force cali-

bration laboratories in the United States and Guam. Ebbert vacationed in Salt Lake City in 1971, visited Stabro, and was "hired on the spot. We are always looking for people with his ability," states Reber. Ebbert began at Stabro as a technician, became a supervisor, and then general manager.

"What is important in our business is traceability," says Reber. "Everyone needs to be able to provide traceability back to the National Institute of Standards and Technology, and that is the service that we provide for our customers."

There are two levels of calibration at Stabro: The primary lab houses equipment that provides direct traceability to NIST; the secondary lab provides traceability through the primary lab and is used to certify equipment that can function with less accuracy.

Stabro also provides its customers with complete documentation of all tests that are performed. The computerized system has information on more than 20,000 pieces of equipment. "We provide our customers with a report card on how their equipment has fared over a number of years. That way they can judge its reliability," says Ebbert.

Stabro calibrates and maintains everything from torque wrenches to sophisticated electronic equipment. For more than 600 regular customers, Stabro Laboratories, Inc., sets the standard.

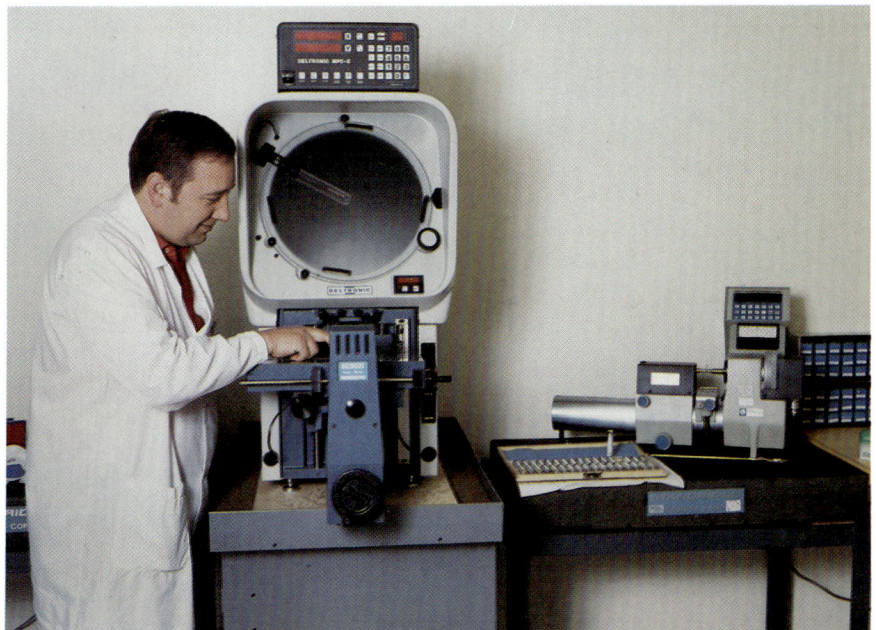

DESERET MEDICAL, INC., BECTON DICKINSON

Deseret Medical, Inc., a wholly owned subsidiary of Becton Dickinson, is a market leader of disposable medical, surgical, and vascular access products, which are sold on a worldwide basis.

In 1956 Deseret Pharmaceutical began as a locally owned, regional distributor of pharmaceutical products manufactured by others. The following year the company was assigned the rights to a patented intravenous (IV) catheter, an event that changed the future of the firm. The catheter was one of the first sterile, single-use disposable devices available to the medical profession. It was unique because it could be placed directly into a vein without the need for gloved hands and a complicated "cut down" surgical procedure.

Deseret dropped the pharmaceutical lines in the mid-1960s to concentrate its efforts on disposable medical and surgical products. Technological advances in intravenous catheter devices and other products established Deseret as a pioneer in the field and a market leader. The com-

pany's success continued through the 1970s, building Deseret's reputation for high-quality, low-cost products that contribute to improved patient health care worldwide.

Warner-Lambert Company acquired Deseret for $130 million in 1977 and held it until 1986, when it was acquired by Becton Dickinson and Company for $225 million.

Becton Dickinson is a major manufacturer of health care and industrial safety products. These range from single-use needles and syringes, safety gloves, and clinical thermometers to highly complex, state-of-the-art chemical and electronic systems for advanced research, diagnosis, and treatment in the field of medicine and its allied sciences.

Deseret built the original Sandy plant in 1967, and has expanded through the years to its current size of 500,000 square feet. Further expansion occurred from 1982 to 1984 to include a 28,000-square-foot marketing administration building; a 34,000-square-foot, state-of-

the-art research and development building; and an off-site, 112,000-square-foot finished goods warehouse and distribution center.

With approximately 1,500 people, Deseret Medical, Inc., is considered a major employer in the Salt Lake Valley. About 1,000 employees are direct labor, or hourly workers, who make the products, and two-thirds of these people are women. Approximately 35 percent are from minority groups, and within this minority group, 73 percent are Asians. There are also about 80 field sales representatives located throughout the United States.

Deseret Medical, Inc., is proud to be a world market leader in high-quality, low-cost products. The firm is committed to providing quality products that reflect a high concern for safety and ease of use to medical professionals worldwide.

Located in Sandy, Deseret Medical, Inc., is one of Salt Lake Valley's largest employers.

EASTMAN CHRISTENSEN

Although Salt Lake City is not recognized as an oil center, it is the location of the corporate headquarters for one of the most successful oil field service companies in the world.

Eastman Christensen is deeply involved in the global search for petroleum and gas deposits. The firm is the world's leading supplier of technologically advanced petroleum drilling products, services, and systems that improve drilling efficiency. Customers include national and international oil companies, independents, and drilling contractors in major oil field centers worldwide in such places as the North Sea, the Far East, Alaska, the Gulf of Mexico, South America, and the Middle East.

In order to serve these customers the firm has more than 2,000 employees worldwide. It also operates 40 facilities in 20 countries, with manufacturing capabilities in five countries—the United States, Canada, Singapore, West Germany, and France.

Eastman Christensen is a new joint-venture between Texas Eastern Corporation (based in Houston, Texas) and Norton Company (based in Worcester, Massachusetts). It combines two trusted oil field service companies—Eastman Whipstock, founded in the early 1930s in California, and Christensen Diamond, founded in 1944 in Salt Lake City.

The Eastman business was started by H. John Eastman, who pioneered the development of controlled directional drilling and well surveying. He was also instrumental in developing techniques for drilling relief wells for blowouts.

Christensen Diamond was founded by Frank and George Christensen (unrelated) who met while playing professional football for the Detroit Lions during the 1930s. After World War II they started the company in Salt Lake City principally to manufacture diamond drilling bits for the western mining market.

Today Eastman Christensen is broadly recognized throughout the oil industry for its quality drilling products and services. These include diamond drilling bits, high-performance drilling systems, downhole motors and related tools, survey equipment, coring services, lateral drill-

ing, and directional drilling services.

At its Salt Lake City headquarters, Eastman Christensen has more than 300 employees working at three different locations. A world-class materials research center is also located in southwest Salt Lake. At this facility sophisticated scientific equipment is used to investigate and develop new cutting/drilling materials for use in bits.

Through the years Eastman Christensen has been one of Utah's leading exporters and one of the world's largest consumers of industrial diamonds. The company has also been actively involved in support of the Salt Lake community through youth programs, United Way, the arts, company-sponsored volunteer programs, and many related activities.

With annual sales in excess of $250 million, Eastman Christensen is a major supplier to a worldwide industry and a major contributor to the Salt Lake business community.

RIGHT: A research technician at the company's Salt Lake City Diamond Technology Center analyzes core samples cut with a diamond bit.

BELOW: Eastman Christensen serves the worldwide petroleum industry with a broad package of technologically advanced products and services that enhance drilling efficiency.

CHAPTER ELEVEN

BUSINESS AND PROFESSIONS

T he Salt Lake Valley's business and professional community brings a wealth of service, ability, insight, and development into the area.

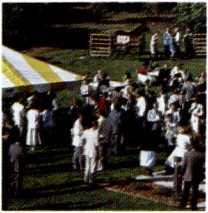

Salt Lake Area Chamber of Commerce, 166

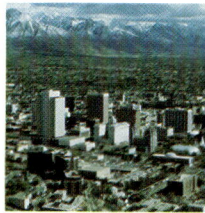

Utah Economic Development Corporation, 167

Arthur Andersen & Co., 168

Johnson & Higgins, 169

Snow, Christensen & Martineau, 170-171

IBM, 172-173

3M Health Information Systems, 174-175

Van Cott, Bagley, Cornwall & McCarthy, 176

Photo by Steve Greenwood

Jones, Waldo, Holbrook
& McDonough, P.C., 177

Murdock Travel Manage-
ment, 178-179

Peat Marwick Main &
Co., 180

Babcock Pace + Asso-
ciates, Architects, 181

City Centre/A Price-
Prowswood Development,
182-183

SALT LAKE AREA CHAMBER OF COMMERCE

Since its inception in 1902, the Salt Lake Area Chamber of Commerce has served as the voice of the business community. It is a private, nonprofit organization whose mission is to promote mutually beneficial relationships among Chamber members and between the community and Chamber members.

The Chamber has always been at the forefront of the community's critical concerns. Through the years many major issues such as the removal of radioactive Vitro tailings and a repayment plan for the Central Utah Water Project were accomplished with the Chamber's help.

Establishing a climate where business can flourish continues to be a major focus. In 1987 the Chamber joined with public and private economic development agencies in Salt Lake County to form a unified entity, the Utah Economic Development Corporation, to pursue new businesses and jobs.

The Chamber's many committees and activities provide its members with involvement opportunities that benefit the volunteer and his/her business while helping the community and the Chamber. These functions fall under the jurisdiction of five councils, each with a specific

The Business After Hours mixer (shown here at University Park Hotel) is one of many networking opportunities the Chamber provides to its members.

mission.

The Special Programs Council's twofold purpose is to make available increased opportunities for members to establish business contacts and promote their companies through Business After Hours membership mixers and the Business to Business Expo. It also provides educational opportunities for Chamber members and other targeted groups through Leadership Utah, the Women & Business Conference, Utah Business Week, and other special functions.

The Small Business/Community Development Council serves as a resource for small businesspeople and offers programs to assist them in running their companies. It also helps improve the area's quality of life with involvement in areas such as downtown revitalization and tourism.

Acquiring and maintaining sufficient funding for the Chamber's overall operation through membership recruitment and retention is the objective of the Membership Council.

Government Affairs' mission is to interact with elected and appointed government officials and to involve members in government relations and the lobbying process. The Communications Council must inform members and the community about the Chamber's workings.

With the help of countless dedicated volunteers the Chamber's staff and 25-member Board of Governors will maintain the course begun in 1902. The Salt Lake Area Chamber of Commerce will continue to be the powerful mouthpiece for the area's business interests in a rapidly growing global economy.

The George S. Eccles Board of Governors Room, located at the Chamber of Commerce Building, hosts many of the area's important business meetings.

UTAH ECONOMIC DEVELOPMENT CORPORATION

State and local government entities in Utah have long given the promotion of economic development high priority because of its promise of new jobs, a broader tax base, and other benefits.

During the past decade dozens of organizations were formed for the express purpose of attracting new businesses, and their efforts produced notable results. However, in the mid-1980s business leaders realized their small development organizations were competing not only with other states for business but with each other as well. Leaders from both the public and private sectors came to the decision economic development had to be done differently. Too many entities were competing with too few resources for new business.

In an effort to combine all major economic development activities in Salt Lake County, the nonprofit Utah Economic Development Corporation (UEDC) was formed in August 1987. UEDC is an unprecedented coalition of public and private individuals and groups who recognize that greater economic development holds the key to Utah's continued growth and development.

"In order for Utah to continue to make its voice heard in a highly competitive national and international arena, we must adopt a new, unified, cooperative approach that stresses communication and coordination, not internal competition," says Nick Rose, chairman of the board of UEDC. "We must concentrate our financial and human resources to produce greater impact than we can achieve working independent of, and, sometimes, against each other."

UEDC's 35-member board of trustees includes the 12 mayors in Salt Lake County, the three Salt Lake County commissioners, 16 leaders of business and industry representing the private sector, and three community-at-large members.

"We have a dynamic story to tell potential investors in Utah," says Rick

Thrasher, president and chief executive officer of UEDC. "Utah has high worker productivity, competitive wage rates, attractive cost of operations, an extremely well-educated work force, centrality to the western part of the country, reasonable-cost land and buildings, and a real sense of community."

"Utah offers a very attractive way of life," he adds. "And UEDC is getting the message out. In the past year (1988) we've been involved in the creation of several hundred new jobs, and in the relocation to Utah of five new companies."

Thrasher has spearheaded successful economic development campaigns in Indiana, Florida, and Wisconsin. He holds the Certified Economic Developer designation from the American Economic Development Council and serves on its board of directors. He also is vice-chairman of the council's certification board.

Perhaps UEDC's most visible achievement to date is its "Utah. A Pretty, Great State" campaign. The project was designed to emphasize positive attitudes and perceptions among Utahns about their state. Print and television advertisements promoting Utah, and billboards splashed with the "Pretty, Great State" logo were seen by hundreds of thousands of Utahns. "The campaign was tremendously successful," says Rose. "Some people liked it, some didn't, but it made for some great dialogue among Utahns

Rick Thrasher, president and chief executive officer of Utah Economic Development Corporation.

about the reasons they like living here." UEDC hopes those positive attitudes will translate into positive images to outsiders.

According to Rose, "In Utah our best sales tool is our state and the people who call it home."

Says Thrasher, "Our challenge at UEDC is to tell everyone else just how great it is to live here."

Utah Economic Development Corporation combines all major economic development activities in greater Salt Lake City, contributing to the growth the area has seen in the recent years.

ARTHUR ANDERSEN & CO.

"Figures that prophesy mean the success or failure of men in business. Any business-man who looks upon his accounting as mere recording and not as a method of control . . . misses the vital significance and use of the facts behind the figures."

So said Arthur Andersen in 1913 when he began his innovative and then-revolutionary financial services organization. Andersen was a visionary, and the first to realize that the traditional bookkeeping of his day could be expanded to forecast the future economic growth of an individual business. Andersen was also the first to hire college graduates, "people who know how to think," as full-time professionals, rather than the part-time schoolteachers used by other firms. His practices were unheard of at the time, and deeply unsettling to the economic community. One competitor predicted Andersen would not only create his own demise, but would destroy the entire profession of accounting as well.

Seventy-five years later the accounting profession is doing quite well, and Arthur Andersen & Co. is one of the world's leading professional firms. With 2,000 partners in every major city and services spanning the economic spectrum, Arthur Andersen's commitment remains the same as ever: to focus the firm's entire resources individually on the clients it serves.

Arthur Andersen's Salt Lake City office is well established and widely known in

the Salt Lake community. In addition to the traditional auditing, accounting, and tax services, Arthur Andersen serves the Salt Lake Valley's financial needs in a variety of other areas, examples of which are personal financial planning, appraisals, litigation support, management information services, customs duties, property tax issues, and employee compensation planning.

While well known for the services it pro-

vides to large multinational companies, Arthur Andersen's Salt Lake office specializes in small business. As office managing partner Dallas Bradford says, "There is not an element of the community, small or large, that we do not serve."

Arthur Andersen is deeply involved in the Salt Lake community. Its employees are active as volunteers in the Chamber of Commerce, the Utah Symphony, United Way, Junior Achievement, Pioneer Theatre Company, the Snowbird Institute, and the Utah Arts Festival, among many others. Management at Arthur Andersen believes in a philosophy of partnership with the community; it depends on the community for its growth, and gives back its own resources of time and commitment. Arthur Andersen & Co. looks forward to the future progress of the Salt Lake Valley, and to the challenge it is going to meet in serving that growth.

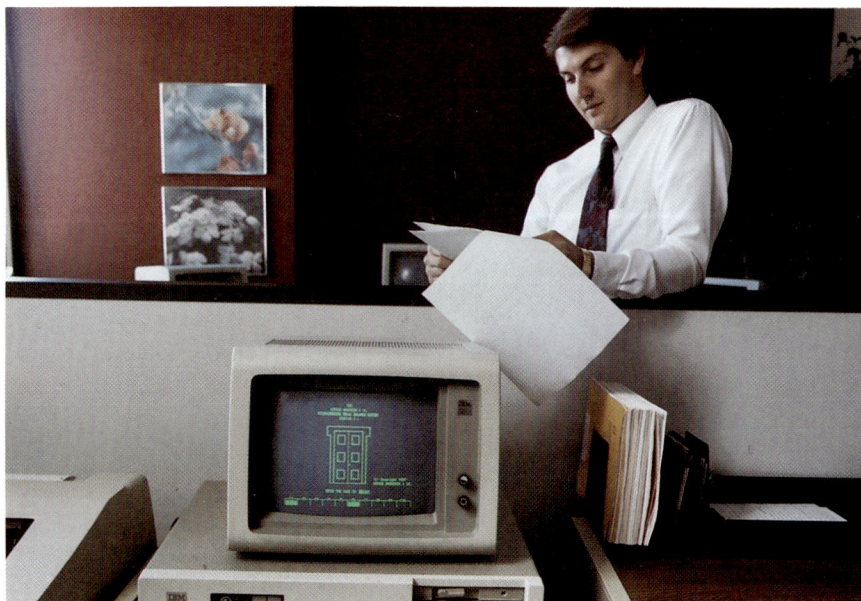

TOP: Arthur Andersen & Co. members confer in the firm's boardroom.

LEFT: Arthur Andersen's employees actively utilize computer technology to offer a wide spectrum of financial services.

JOHNSON & HIGGINS

Johnson & Higgins is the world's largest privately held insurance broker and employee benefits consultant. In 1845, even as wagon trains were opening up the West, the partnership of Henry W. Johnson and A. Foster Higgins was formed to handle marine average adjusting and coverages for shipping companies. In its early years the firm handled such exciting events as settling losses after the catastrophic San Francisco earthquake, and placing insurance on the *Titanic.* (When the ship sank, every claim was handled within 30 days.)

In the latter part of the twentieth century, almost 145 years after its inception, Johnson & Higgins has matched its clients' evolution into an expansive international organization. What began as a two-man operation in New York has become a 7,000-employee organization with more than 100 offices worldwide. More than

700 of the *Fortune* 1,000 companies use J&H for all or part of their insurance placement, risk management, and consulting needs. Through the years the firm has developed expertise in areas undreamed of by the founders, such as captive management, loss control, executive compensation consulting, design of self-insurance and risk-retention programs, and actuarial projections.

In 1986 Johnson & Higgins opened its 46th United States office in Salt Lake Valley. Its goal was to bring to the West the sophisticated risk management techniques used successfully in other parts of the country. As Utah-based companies have matured, their need for local access to national and international professional assistance in protecting corporate assets and managing employee benefit programs has increased. The success of Johnson & Higgins in meeting this need is

evident in the growth of the Utah office from two professionals three years ago to a staff of more than 20 in 1989.

Located in the Eagle Gate Plaza, the local Johnson & Higgins team, along with its national and international counterparts, is dedicated to providing superior service to a wide array of clients. Whether a bank or a contractor, an industrial giant or a small, start-up firm, Utah organizations will find the same competence and integrity in Johnson & Higgins that has characterized the firm for more than 140 years.

LEFT: Daniel L. Jones, senior vice-president and branch manager.

RIGHT: David C. Heslington, vice-president and large account manager.

SNOW, CHRISTENSEN & MARTINEAU

Snow, Christensen & Martineau, one of Salt Lake City's largest law firms, traces its history to the year 1878, when it was founded in Provo, Utah, by Samuel R. Thurman. In 1886 Thurman formed a partnership with George Sutherland, who later served on the Supreme Court of the United States. In 1906 the firm moved to Salt Lake City as Thurman, Wedgwood & Irvine. The practice became Worsley, Snow & Christensen in 1967, and nine years later adopted its present name, Snow, Christensen & Martineau. In addition to Justice Sutherland, past firm members include two Utah Supreme Court justices, a United States senator, a deputy attorney general of the United States, and a dean of the University of Utah College of Law.

Snow, Christensen & Martineau has grown significantly in recent years, and is committed to a policy of continued growth so long as the quality of its clientele, personnel, work product, and working environment can be maintained. The firm takes pride in its history and in its tradition of high-quality legal work and professional integrity.

In the 1950s the firm's principal areas of practice were commercial and insurance defense litigation, and corporate and motor carrier law. During that decade the business department expanded with the Colorado Plateau uranium boom.

During the 1960s a conscious decision was made to develop expertise in other areas of the law through active recruiting of experienced attorneys as well as recent law school graduates. In 1965 Utah adopted the Governmental Immunity Act, and the firm became defense counsel to cities, counties, boards, and governmental agencies statewide.

Rapid expansion of commercial litigation came in the 1970s as large and complex antitrust and securities litigation became increasingly commonplace. At the same time, the real estate department grew to meet the needs of local and regional developers of shopping centers, apartment buildings, condominiums, and commercial structures. The litigation department also expanded to meet the growth in the products liability area and in medical malpractice defense. The tax and employee benefit department grew with the increased number of business clients.

In the 1980s Snow, Christensen & Martineau has added natural resources, energy, municipal law and financing, and telecommunications law. The firm has also developed a large and diverse health law practice.

A substantial part of the exceptional growth of Snow, Christensen & Martineau is due to the attraction of new clients, who believe that the unique combination of strong litigation capability, business expertise, and substantive law competence will best serve their needs.

The firm has long recognized its responsibility to the legal profession and to the

Snow, Christensen & Martineau, a law firm established in 1878, continues to serve local and national clients from its offices at 10 Exchange Place.

The Newhouse Building, with its marbled lobby and neoclassical structure, is the home of the law firm of Snow, Christensen & Martineau and its 160 employees.

ees of Utah State University; president of the Salt Lake City Legal Aid Association; member of the boards of the Utah Zoological Society, Salt Lake Art Center, Ballet West, and City Landmarks Commission; and president of Big Brothers-Big Sisters of Greater Salt Lake.

Snow, Christensen & Martineau is a strong supporter of the Utah State Bar Association and its activities. Recently three of the firm's members have been president of the Utah State Bar, and five have been president of the Salt Lake County Bar Association. The firm's lawyers have been instrumental in founding and developing the nationally prominent Inns of Court Program to train young lawyers, and are actively involved in all levels of bar activities. Currently four members of the firm have been elected to the prestigious American College of Trial Lawyers.

Beginning in the early 1960s Snow, Christensen & Martineau made a commitment to collect and decorate its offices with the work of early Utah artists. Today the offices contain perhaps the largest private collection of early Utah art in the state. The collection includes works by Cyrus E. Dallin, John Hafen, Mahonri Young, J.T. Harwood, C.C.A. Christensen, George M. Ottinger, A.B. Wright, and Waldo Midgley.

In 1981 shareholders of the firm purchased the Newhouse Building, located at 10 Exchange Place in the historic financial district of Salt Lake City. This building was part of the dream of Samuel Newhouse, a turn-of-the-century mining magnate, who desired to erect a Wall Street of the West. The building is listed on both the national and state historic registers. The firm's restoration of the top floor where Samuel Newhouse had his private offices earned it the Award of Merit from the Utah Heritage Foundation. Snow, Christensen & Martineau currently occupies 60 percent of the building, while other business and professional oper-

ations lease the remaining space.

As legal markets have grown and shifted, the firm has responded by diversifying to become a full-service law firm. Its attorneys practice in virtually every field of law, including corporate and business, insurance defense (with large sections in products liability and medical malpractice defense), energy and natural resources, general litigation, employee benefits, tax and probate, real estate, municipal law, and finance and health law.

As Salt Lake City and the Intermountain region continue to grow and develop through new service and business companies, Snow, Christensen & Martineau continues to grow and develop its expertise in those areas of law that it believes will be critical to the future of those companies and to the future of the Intermountain region.

community. Within the past several decades its attorneys have made important contributions to the Intermountain region in many areas, including roles as president of the International Association of Lions Clubs; consultant to the American Delegation to the United Nations Peace Conference; president and trustee of the Utah State Training School; board of trust-

IBM

International Business Machines Corporation has been an integral part of the Salt Lake City business community for 60 years. From a very simple beginning, with only a few employees selling time clocks and punch card equipment to local companies and distributors, IBM has become a major Utah employer serving small and large businesses in all industry sectors along the Wasatch Front.

In IBM's five-story, 130,000-square-foot building at 420 East South Temple, more than 400 employees are concentrating on

solutions to the valley's need for advanced information processing. As elsewhere, Salt Lake City's business, industry, and public sector require new products and technologies for computer processing, which has become a fundamental part of virtually every organization's daily operations. IBM and its people are dedicated to maintaining the company's solid reputation for providing quality information, systems products, and support. The primary marketing emphasis today, however, is on developing application solutions that meet specific customer needs, with the hardware platform being a secondary consideration.

Another key ingredient in Salt Lake City's future is the requirement for further application of information processing to office productivity and engineering/scientific advancement. IBM marketing, systems engineering, and service representatives are working to bring to Wasatch Front customers the latest in products and services for improving and streamlining a broad variety of office activities, from financial analysis and word processing to graphics and printing. With its

advanced-function workstations, supercomputers, and artificial intelligence, IBM is also addressing the increasing need for greater capacity and sophistication in engineering/scientific and industrial computing. Telecommunications is also another rapidly growing area that helps integrate new technologies in a changing environment.

IBM's successful association with Salt Lake City is part of a long company history that dates back to 1917 when its predecessor, the Computing-Tabulating-Recording Company, had only a few employees selling time-recording equipment and computing scales. C-T-R had been formed in 1911 by a merger of three operations that included the Tabulating Machine Company (founded in 1896 by Herman Hollerith, the father of modern data processing) and the International Time Recording Company of Endicott, New York (the site of IBM's oldest manufacturing and development facility).

In 1924, under the able leadership of Thomas J. Watson, Sr., the firm changed its name to International Business Machines Corporation. Through a combination of his own moral fervor, a concern for his employees, and salesmanship, Watson created a corporate culture and an esprit de corps that remains the envy of the business world today. He also conceived such expressions as "There's no saturation point in education," and the corporate slogan for many years, "Think."

By 1935 the company's product line had expanded to include electronic accounting machines and typewriters—the forerunners of today's productivity tools. By the late 1950s its systems had evolved to include the early prototypes of modern-day computers.

Today IBM is a corporation of about 400,000 employees in more than 130 countries, developing, manufacturing, marketing, and servicing computer systems ranging from large-scale processors to mid-range systems to the IBM Personal Computer. Also included are application software, robotics, industrial systems, education, and a broad range of professional services. Respect for the individual, delivering outstanding customer service, and a commitment to excellence remain IBM's basic beliefs.

TOP: IBM punched-card calculating in the past.

ABOVE: IBM's Salt Lake City branch office.

LEFT: People—dedication to excellence.

FACING PAGE, RIGHT: An IBM marketing team planning meeting.

Utah IBM people and their counterparts worldwide have benefited from a number of company programs. The firm is an equal opportunity employer, ensuring that all employees have the same chance to succeed without regard to race, color, religion, sex, national origin, handicap, or age. Hiring is based on business need, job-related requirements, and an individual's qualifications; advancement depends on job performance and the individual's demonstrated ability to assume greater responsibility. Affirmative action programs also are in place to ensure that all can compete on an equal basis. A merit system of pay and promotion rewards superior performance. With IBM's full employment policy, no person employed on a regular basis for the past 50 years has lost as much as one hour of working time because of a layoff.

IBM, a member of the *Fortune* 500, is recognized as a leading world corporation today. As recently as 1987 the company was named "Number One in the World for Quality," and the Presidential Commission on Employment for the Handicapped recognized IBM as Employer of the Year.

Wherever it does business, IBM contributes significantly in funds, equipment, and employee services to support education, social programs, and the arts. In Salt Lake City, the firm donates in excess of $400,000 annually to local non-profit organizations, including United Way, Ballet West, the Utah Heart Association, the Utah Special Olympics, and various job training groups. (This amount,

ABOVE: A marketing representative on-line to the IBM network.

along with other direct expenditures in the community, generates an economic multiplier impact on the state of more than $25 million annually.) Salt Lake City IBM managers and employees volunteer their time to such diverse interests as the State of Utah Education and Economic Development Task Force, the Utah Manufacturers' Association, the Salt Lake Area Chamber of Commerce, the United Way, Boy Scouts, the Red Cross, and local civic organizations. These activities illustrate the fact that IBM and its people have always felt a responsibility to leadership and civic participation by supporting educational, cultural, and social programs in the communities where its employees live and work.

IBM's partnership with Utah businesses and public sector agencies is based on commitment, dedication, excellence, and good will. This association is the foundation from which IBM employees strive together in forging new directions and a better life for their company, the Wasatch Front, and its people.

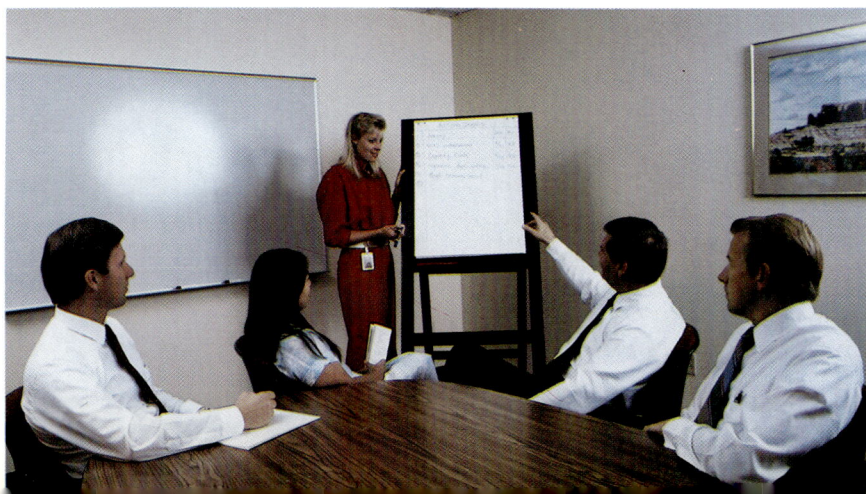

173

3M HEALTH INFORMATION SYSTEMS

3M Health Information Systems (HIS) is an integral part of daily life in a hospital. Its software computer systems save countless hours of time for medical personnel, and its quick access to patient information has even been credited with saving lives.

3M HIS allows medical staff to keep track of all aspects of individual patient care, is invaluable in hospital laboratories to process tests and other procedures, ensures optimized medicare reimbursement, and allows rapid access to past medical records.

One of the leading suppliers of hospital computer systems in the country, 3M HIS is also one of the largest companies in Utah. Its Salt Lake-based employees develop new software, market the product, provide technical support, and do accounting and secretarial work. 3M HIS employees nationwide sell and service the finished products.

Salt Lake City's 3M HIS is a division of 3M, based in St. Paul, Minnesota. 3M

HIS belongs to 3M's multibillion-dollar Life Sciences Sector and enjoys the benefits from constant access to the full spectrum of 3M's technologies and expertise. It is said that nearly half the world's population benefits daily in some way from one or more of 3M's products. 3M is one of the fastest-growing providers of health care products in the world, and also one of the most diversified. 3M's goal is to achieve 25 percent of sales each year from new products that did not exist five years prior. It is this spirit of entrepreneurial growth that set the stage for the emergence of 3M HIS.

In the early 1980s changing Medicare and health care requirements on both

RIGHT: A.E. Eggert, general manager of 3M Health Information Systems.

BELOW: This artist's drawing depicts the 90,000-square-foot facility for 3M Health Information Systems.

the state and federal level established the need for rapid computerization within the hospital industry. 3M was already well established in the hospitals, having sold medical record storage products—primarily micrographic equipment and

supplies—plus a wide variety of products for health care professionals and patients.

This in-depth knowledge of hospital information requirements, combined with a corporate culture that encourages development of innovative new products, convinced 3M that it had the right resources to develop a broad spectrum of information products that would help hospitals meet ever-changing government requirements, ensure quality of patient care, and improve productivity.

Dr. John Morgan is credited with 3M HIS's Salt Lake City location. Morgan, a medical biophysicist and former University of Utah professor, is one of the pioneers in the hospital computer field. By the early 1980s, with 27 employees, Morgan had developed and was successfully marketing a system called Code 3™. 3M believed Code 3™ could be the cornerstone of its new division, and in 1983 the system and 3M successfully merged. Today both Code 3™ and Morgan remain important components of the 3M HIS business.

"In the past five years our customer base has increased fourfold to a total of 1,200," says Al Eggert, 3M HIS general manager. "I anticipate that the hospital information systems market will continue to grow at an annual rate of more than 12 percent. We are looking forward to the rapid growth of 3M HIS to more than 500 employees in the near future."

To accommodate 3M HIS' rise, Eggert recently moved his staff into a new multimillion-dollar facility in the south end of the Salt Lake Valley. The modern, 90,000-square-foot office solidifies the company's Salt Lake location and brings all of 3M HIS' operations under one roof.

3M HIS consists of three major product areas: Code 3™, H.E.L.P.™ patient care, and Medlab™ Laboratory Systems. Code 3 is a set of software systems for clinical data management that helps medical record employees to code more accurately, consistently, and completely. Code 3's Codefinder™-brand software and DRGfinder™ software are two of the early systems that proved computers could be a powerful tool for ensuring optimum Medicare reimbursement. More recently HCPCS/CPTfinder software has

enjoyed especially rapid growth because it enables hospitals to meet the new HCFA regulation that requires coding of ambulatory care information. Code 3™'s research and development team continually works to meet the changing needs of the medical record industry.

Building on its strengths in the medical record department, 3M HIS continued to expand its presence in the multibillion-dollar turnkey hospital systems industry. To that end, it acquired Control Data's Hospital Information Systems business in 1987. This business consists of H.E.L.P. comprehensive patient care system, which provides information for patient care and financial management, and Medlab™ Laboratory Systems. Medlab products are designed to ensure quick, concise reporting of laboratory tests.

3M believes that this new organization will lead to the enhancement of existing products and pave the way for new ones. "Our goal," says Al Eggert, "is to provide health care professionals with decision, consultation, and information presentation tools that support high-quality, efficient patient care while meeting the administrative and financial needs of the hospital.

"We also believe in being good corpo-

3M Health Information Systems is among the market leaders as a provider of customized hospital computer systems.

rate citizens," says Eggert. "We encourage our employees and our board to participate on a volunteer basis in support of the arts and Salt Lake's nonprofit agencies."

3M HIS benefits from interaction with local medical facilities. "We have strong working relationships with Intermountain Health Care and the University of Utah Medical School and Hospital. We hire many of our employees from the excellent Medical Informatics Department at the University of Utah. We also work closely with LDS Hospital. Some of our programs were developed there, as well as at the U," says Eggert.

3M HIS' powerful corporate backing and its cutting-edge computer systems ensure its continued success. "We are looking to be the top provider of clinical information systems to the health care marketplace," says Eggert. "We look forward to 3M's association with the Salt Lake business community, and to the growth of our operations there."

VAN COTT, BAGLEY, CORNWALL & MCCARTHY

Van Cott, Bagley, Cornwall & McCarthy maintains its principal offices in Salt Lake City and a branch office in nearby Ogden. It is the largest law firm in Utah and is one of the largest in the Intermountain West.

Van Cott lawyers have been prominent in the legal community since the firm's founding in 1874. Among its members and alumni are former state and federal district court judges, past presidents of the Utah State Bar and Salt Lake County Bar Associations, and the late George Sutherland, who during his career in public service served as a congressman, a senator, and an associate justice of the Supreme Court of the United States.

Van Cott is a professional corporation, governed by a board of directors and managed in its daily affairs by a management committee. The firm's attorneys are organized into five practice sections in order to deliver specialized services: banking, finance, and creditors' rights; corporate and commercial; litigation; natural resources; and tax, probate, and estate planning.

A full-service law firm, Van Cott represents an impressive range of commercial, industrial, and financial clients in diverse sectors of the economy. A significant portion of Van Cott's practice, however, is devoted to providing legal services directly to individuals, particularly in the areas of tax, probate, and estate planning. Van Cott lawyers also provide pro bono legal services to individuals and nonprofit organizations.

In addition to legal work, Van Cott, Bagley, Cornwall & McCarthy attorneys dedicate themselves to public service, volunteering their time and expertise to such diverse groups as the Utah Symphony, Ballet West, United Way, Traveler's Aid, and the Legal Center for the Handicapped. The firm is especially proud of its Mentor Program in the public schools. The goal of the Mentor Program is to provide information to students about the legal system and to make students more aware of their rights and responsibilities as citizens. On a rotating basis, lawyers in different areas of expertise teach a civics class in Salt Lake-area high schools. The firm also hosts an annual career day at its Salt Lake City offices to introduce students to the many career opportunities available in the legal field.

RIGHT: The Van Cott management committee members are (from left) Stephen D. Swindle, David E. Salisbury, and Robert D. Merrill.

BELOW: Van Cott attorneys David L. Gillette and Kathryn Snedaker pause in the firm's Salt Lake City reception area to confer about their client's request for legal opinion.

JONES, WALDO, HOLBROOK & MCDONOUGH, P.C.

Jones, Waldo, Holbrook & McDonough, P.C., is one of the largest and fastest growing legal firms in the Intermountain West. With offices located in Salt Lake City and St. George, Utah, and Washington, D.C., it provides full-service legal work for large and small businesses and individuals in the region and throughout the nation.

The firm traces its roots back to the year 1875 and Joseph L. Rawlins, Utah's first full-term senator and the founding father of Jones, Waldo, Holbrook & McDonough.

From its beginning Jones, Waldo, Holbrook & McDonough, P.C., has played a prominent role in national and local affairs. Members of the firm perform leading roles in political, cultural, civic, and legal life. Lawyers still active in the firm have held present or past positions as governor of Utah, chairman of the Board of Regents, chairman of the State Board of Education, chairman of the University of Utah Institutional Council, member of the Utah State Senate, members of the boards of directors of major corporations, bar commissioner, and bar presidents. Firm members contribute both time and expertise to their community by supporting various nonprofit agencies, and by serving on the boards of directors of such organizations as the Utah Symphony, Ballet West, the Utah Opera Company, and the Salt Lake Art Center.

Lawyers in the firm are organized into and concentrate their practice in specific areas to provide the highest-quality legal service. These areas include corporate banking, securities regulation, antitrust, communications, bankruptcy and creditors' rights, employee benefits, hospital and health law, labor relations, trial and appellate litigation, municipal finance, estates and trusts, real estate, utility law, tax, patents, natural resources, media, constitutional law, and agricultural law.

Among the firm's clients are major investor-owned public utility companies that operate in Utah and in a number of surrounding states, a national food and drug chain, a major banking organization, Utah's largest newspaper, agricultural cooperatives, security broker dealers, natural resource companies, and many other corporate enterprises.

The Washington, D.C., office serves the needs of the firm's clients in connection with federal regulatory and congressional matters. Through the Washington office, the firm is active in administrative and communications law and many matters directly affecting the state of Utah.

The St. George office serves southern Utah, a rapidly expanding area that is emerging as one of the Intermountain West's most attractive business environments.

Today Jones, Waldo, Holbrook & McDonough, P.C., is one of the leading law firms with headquarters in the Intermountain West. Its rapid growth, commitment to very high legal ability, and accomplishment-oriented practice continue to attract and maintain a major and diverse business-oriented client base.

ABOVE: Jones, Waldo, Holbrook & McDonough, P.C., with offices in Salt Lake City and St. George, Utah, and Washington, D.C., is recognized for its very high legal ability. Photo by Borge Andersen

LEFT: Jones, Waldo, Holbrook & McDonough, P.C., was a pioneer legal firm in the Salt Lake Valley. From left are partners Donald B. Holbrook, Kay S. Cornaby, Calvin L. Rampton, W. Robert Wright, and Randon W. Wilson. Photo by Borge Andersen

MURDOCK TRAVEL MANAGEMENT

Murdock Travel Management is not a typical travel agency; in fact, Murdock may be the only travel agency in the world that can trace its founding to a church assignment. Today Murdock Travel is one of the largest agencies in the United States specializing in foreign travel. It employs more than 240 travel specialists in 16 offices worldwide. In 1936, when Franklin J. Murdock set out for Holland as a mission president for the LDS (Mormon) Church, little did he realize that his church calling would subsequently lead to a worldwide business.

Franklin Murdock's mission call coincided with the outset World War II. In 1939 LDS leaders in Utah, driven by a growing concern for their church workers as the war escalated, asked Franklin Murdock to assist in the immense job of bringing home the hundreds of missionaries and support personnel from Europe. Murdock then arranged the safe exit of church volunteers serving on the European continent, a heroic task that included mountains of international documents and the unique ability to cut "red tape," while successfully maneuvering through the diplomatic channels of a dozen European countries in the midst of war.

Franklin Murdock carried out his assignment with such finesse that after his return to Utah he was offered the position of church travel secretary. For the next 18 years Murdock personally handled all of the travel arrangements for LDS missionaries worldwide, as well as the itineraries of all church authorities. By 1958 his responsibilities had escalated to the point where president David O. McKay believed he could better serve the church's needs as a private entity. Murdock Travel Agency was formed to provide full travel services to the LDS Church and the private sector.

Thirty-one years later Murdock Travel Management has expanded to become a recognized leader in professional travel management. Though Murdock Travel Management continues to serve the travel needs of the LDS Church, it also handles such major accounts as Intermountain Health Care, Kennecott Utah Copper, O.C. Tanner, Smith's Management, Pyke Manufacturing, Brigham

Murdock Travel's principal offices occupy space on two floors of the Beneficial Towers. Murdock's Cruise Corner, the Jensen-Baron wholesale travel division, and the Murdock Shuttle are located in the adjacent Eagle Gate Plaza Tower.

Young University, the State of Utah, Eyring Research, Zion's First National Bank, Peat Marwick & Main, Utah Power & Light, Price Savers, Ryder Trucks, Rockwell International, Rocky Mountain Helicopter, Help-U-Sell, Dynix, and the University of Utah. More than 70 percent of Murdock's volume comes from corporate clients.

Murdock president F. Wayne Chamberlain says, "We have solved visa and passport problems, legal questions, and monetary exchanges for Mormon missionaries in so many countries for so many years that we are uniquely well equipped to handle the needs of all foreign travelers." Murdock's regular scheduled cruises and tours include such destinations as Latin America, Europe, the Orient, the Caribbean, and Hawaii.

Panorama Tours, Murdock's "South of the Border" division, and Jensen-Baron (wholesaler-consolidator division) are both growing at a rapid rate. "We're able to specialize in Mexico, Latin America, South America, and the Orient because

of our vast experience," says Chamberlain.

Chamberlain points to his experienced staff as the reason for Murdock's continued success. "We have a unique team spirit. Murdock is a privately owned company whose stock is owned by the employees and the board. More than 40 percent of our employees are graduates of our own travel school. We train the best travel agents in the country."

Chamberlain is also proud of his 38 employees who have achieved the title of Certified Travel Counselor (CTC). "That's comparable to being a CPA in the travel industry. It requires five years of experience and completion of a two-year intensive course. These people have truly

set themselves apart in the industry as professionals. Murdock Travel is the industry leader in professionalism and international experience."

Murdock Travel's directors bring a broad base of knowledge to the company. "Our advantage with our esteemed board is their vast experience in the community. They bring their expertise in economic and social trends to our agency. Their contribution is invaluable," says Chamberlain. Directors include Mack Lawrence, chief executive officer of US West/Utah (Mountain Bell); James Mortimer, publisher of the *Deseret News;* David Hemingway, senior vice-president of Zion's First National Bank; and Wilford Kirton, senior partner in the law firm of Kirton, McConkie & Bushnell.

Longtime employees of Murdock Travel who have helped the agency achieve its current success are its three senior vice-presidents, Garratt T. Beesley, Edwin H. Burgoyne, and Conrad H. Burgoyne, all of whom have been with Murdock Travel for 30 years. Roger G. Stratford, a former Braniff Airline pilot/district sales manager, is vice-president/sales and marketing.

Chamberlain has served as Murdock

Travel's president since 1979. He has extensive business and travel expertise and has served the Salt Lake community by chairing the Associates Committee (all charity care fund raising) for Primary Children's Medical Center since 1981. Additional board service includes Beneficial Life Insurance, Intermountain Health Care Plans, Salt Lake Area Chamber of Commerce, National Cougar Club, the Pioneer Memorial Theatre, Associated Travel Network (International Consortium), Ballet West, the Salt Lake Rotary, and the Salt Lake Convention and Visitors Bureau, among others.

Murdock Travel has offices worldwide in cities ranging from Sydney, Honolulu, and Buenos Aires to Idaho Falls, Rexburg, Ogden, Provo, St. George, and six offices in Salt Lake (Beneficial Tower [three], Holladay-Olympus Cove, and Jensen-Baron in Eagle Gate Plaza [two]). Future plans include four additional foreign offices in the next half-decade and an expansion of services in the Intermountain area. "We will continue to expand out of the necessity to better serve our clients' needs. We expect our growth to be exciting," says Chamberlain.

Franklin J. Murdock served as presi-

dent of his company until age 77. He died in 1986 at the age of 83. His firm has grown from one employee to more than 240 travel specialists. Today Murdock Travel Management carries on his proud tradition of innovative travel with the very highest level of quality customer service and professional travel management.

179

PEAT MARWICK MAIN & CO.

Peat Marwick Main & Co. is the dominant public accounting firm in Utah, its roots well established in the state. It is a well-balanced organization with expertise in carrying out international, corporate, and government assignments, as well as offering a broad range of services to small and mid-size companies.

The Utah office provides expertise, founded in a knowledgeable and experienced staff, in accounting and auditing, tax services, management consulting, litigation support, investigative accounting, bankruptcy, and governmental services.

The staff is specialized along industry lines to stay ahead of the unique aspects of regional and national developments. The firm is strong in finance and financial services, high technology and manufacturing, energy, communications, government, real estate development and construction, litigation support, and the entertainment, restaurant, and recreation field.

With technology constantly changing, Peat Marwick's access to national issues, trends, and data provides the tools necessary to advise clients on all aspects of their business. What is occurring today in London may well impact a Utah client tomorrow.

Peat Marwick Main & Co. was created by the merger of two international accounting firms. Not only did this union create the largest accounting firm in the world, it formed the largest firm in Utah. This merger resulted in a professional service organization with geographical balance, depth, and breadth of services and impressive personnel resources.

The firm is dedicated to the future of Salt Lake City and the State of Utah. Partners, managers, and staff serve in leadership positions in a wide variety of industry, civic, philanthropic, governmental, and arts organizations. These individuals are committed to outstanding client service and improving the community in which they live and work.

From its Salt Lake City offices in the Eagle Gate Tower, Peat Marwick Main & Co. serves its clients in many segments of business, including communications, high technology, and the recreation and entertainment industries. Photo of skiers, courtesy, Snowbird

BABCOCK PACE + ASSOCIATES, ARCHITECTS

Babcock Pace + Associates architects are both masters of innovation and specialists in diversity. The firm's careful, clear planning and attention to detail has created the contemporary City Centre Complex and the Western Institute of Neuropsychiatry, as well as the extensive renovation of the historic Brigham Street Inn and the 100-year-old Trolley Square.

ect is uniquely suited to the owner's needs, site, and resources. Throughout each project, key personnel, including architects John Pace and Fred Babcock and interior designer Deanne Uriona, maintain a close liaison between the design staff, general contractor, and the owner. From the initial stages of planning to the completion of a building's interiors and

renovated, and moved into the historic buildings at 52 Exchange Place. "Our offices had been located at Trolley Square," says Pace, "but by the mid-1980s we were eager for our own building and made a decision to renovate another historic site. We chose the old Newhouse Realty Building for its open-space possibilities, the charm of the area, and because

ABOVE: Western Institute of Neuropsychiatry

TOP LEFT: Babcock Pace + Associates award-winning offices at 52 Exchange Place in Salt Lake City.

LEFT: City Centre

The firm's principals, John E. Pace, A.I.A., and Fred M. Babcock, A.I.A., have years of professional experience in a wide variety of building types. Many of Babcock Pace's projects have received design awards. The firm's regionally and nationally recognized work includes medical facilities, office buildings, historic renovations, research facilities, industrial complexes, multifamily housing, and single-family residences.

Babcock Pace approaches each of its projects with a fresh outlook. The programming, planning, and design for each proj-

landscape, the firm offers its clients individuality coupled with years of experience.

Babcock Pace employs a staff of 17, seven of whom are licensed architects. The firm utilizes some of the latest advances in the architectural design field, Computer Aided Design/Drafting (CADD). CADD affords the designers and clients accurate consideration on a variety of options within limited time restraints.

The firm's commitment to Salt Lake City and the Wasatch Front is exemplified by its own award-winning office space. In 1984 the company purchased,

we wanted to contribute to the vitality of Salt Lake's central business district." Firm members contribute many volunteer hours to the community through the R/UDAT study implementation, the Salt Lake Area Chamber of Commerce, and the Utah Symphony, as well as extensive involvement in national professional organizations.

A widely recognized project is the four-star Brigham Street Inn. Located on East South Temple, this mansion combines the most appealing aspects of both a bed-and-breakfast inn and a luxury executive hotel. Working with A.S.I.D. interior designers, the firm's design gave each of the inn's elegant guest rooms and common areas an individual style and unusual comfort.

Babcock Pace + Associates is instrumental in infusing a distinct personality to the Salt Lake Valley. Perhaps more than any other profession, architecture defines a city's sophistication and livability. The Wasatch Front is fortunate to count Babcock Pace + Associates as one of its premier architectural, planning, and design firms.

181

CITY CENTRE/
A PRICE-PROWSWOOD DEVELOPMENT

City Centre means business.

Once an empty, gutted block in the middle of Salt Lake City, City Centre's elegant office tower is now the cornerstone of Salt Lake's plan to redevelop the southern end of the downtown central business district. Home to the Salt Lake Area Chamber of Commerce and located across the street from the historic City-County Building, City Centre has brought vitality, architectural achievement, and the bustle of commerce back to the heart of Salt Lake Valley.

Designated by early Mormon settlers as "Block 53," the block between Third and Fourth South and State Street and Second East was the site of Tuft's Mansion, the first hotel in the territory, and enjoyed prominence as the city's central business district for many years. But in the late 1960s changes in downtown development patterns took place, and high-rise towers sprung up in other areas as Block 53 fell into disuse.

In 1986, when a multimillion-dollar commitment was made to restore the adjacent City-County Building, city planners saw the potential to return Block 53 to

its rightful place in history as well. They chose Price-Prowswood as its developer, and the result is City Centre, a $100-million planned office/commercial project comprised of luxurious office space, restaurants, and shops, as well as outdoor patios, walkways, and landscaped areas. City Centre is a carefully planned effort designed to frame, without obstructing, the view of the City-County Building to the south, the majestic Wasatch Mountains to the east, and the State Capitol Building to the north.

The first structure at City Centre—the Chamber of Commerce Building/City Centre I—completed in 1986, is a 10-story brick and glass tower housing 208,000 square feet of what developers call the finest office space in the western United States. It includes one level of retail space, nine levels of office space, and two levels of parking below ground. An open promenade at the retail

level receives pedestrians from all sides of the building. Of the 10 levels, the bottom three are low-rise, terraced levels with walkout decks and brick railings. The remaining seven are high-rise levels sheathed in bronze-toned, insulated, reflective glass. City Centre I has been carefully researched to create the most efficient commercial facility in the area while offering maximum adaptability for future change and long-term cost effectiveness.

All offices at City Centre I are custom designed, and amenities include full security, a state-of-the-art telephone system, individual heating and cooling, card access after hours, and two underground levels of well-lit parking facilities. Elegant common areas feature fine carpeting, rich paneling, and commissioned art. Tenants include the Salt Lake Area Chamber of Commerce; Fairchild Communications; Fidelity Investments; Price Waterhouse; the

law firm of Prince, Yeates & Geldzahler; the law firm of Yengich, Rich, & Xaiz & Metos; The City Executive Centre; Stuart James Company; the Utah Economic Development Corporation; Coastal States Energy; the law firm of Kipp & Christian; The Principal Financial Group; the law firm of Brown, Smith and Hanna; Cellular One; Huber, Erickson and Butler CPAs; Consolidated Realty Group; Emnet/Spectrum Engineering; Allen Hospitality; Subway Sandwiches; and the popular social club, Studebaker's. Known locally as the "Chamber of Commerce Building," City Centre I is destined to become the premier business location in Salt Lake City.

The second phase of development, City Centre II, is planned as an architectural accompaniment to City Centre I. Located on the northeast corner of Fourth South and State Street, the 22-story high rise will provide office and retail space of the same high quality and beauty as City

Centre I. Preleasing of office space for City Centre II is brisk, and interest among Class-A office users is high.

City Centre's location is ideal for tenants and visitors who need quick access to city and state government offices. Driveways are designed for easy in-and-out access to the major freeways. Within walking distance is the municipal court system, Salt Lake City Corporation offices, and the federal court system. The Heber M. Wells state office building is located on the northeast corner of Block 53 and houses the state's Tax, Insurance, Industrial, Business Regulations, and Public Utilities commissions. The city's main library and post office are close by, as are dozens of restaurants, movie theaters, social clubs, and retail shops. Several golf courses, Symphony Hall, the Salt Palace, the University of Utah, the University Medical Center, and the Salt Lake International Airport are all within a few minutes' drive.

City Centre offers a combination of the practical and efficient with the elegant and innovative. Whether tenants are inside City Centre looking out at the majestic views or outside enjoying the variety of scenes on the plaza, they will appreciate its ideal setting. Developers call City Centre the most successful office project in Salt Lake City. Its aesthetics, attention to historic surroundings, and roster of top-notch clients ensure its continued success and the return of Block 53 as the commercial and governmental hub of Salt Lake City.

CHAPTER TWELVE

QUALITY OF LIFE

Medical and religious institutions contribute to the quality of life of Salt Lake Valley residents.

Church of Jesus Christ of Latter-day Saints, 186-189

St. Mark's Hospital, 190-191

Photo by Steve Greenwood

CHURCH OF JESUS CHRIST OF LATTER-DAY SAINTS

Salt Lake City was founded in 1847 by early pioneers of The Church of Jesus Christ of Latter-day Saints, but today it is a community of considerable religious and cultural diversity. More than 80 religious denominations are represented in the valley, and local ethnic cultures include Chinese, Greek, Vietnamese, Cambodian, Japanese, Italian, German, Scandinavian, Tongan, and Samoan.

In the midst of this diversity, however, there remain echoes of the city's Mormon pioneer heritage. That heritage is reflected in the prevailing atmosphere of ancestral pride, a strong work ethic, a cooperative community spirit, and, visually, in the pioneer architecture of downtown's historic Temple Square.

Temple Square's 10 acres of historic buildings, visitors' centers, statues, fountains, and lush gardens offer a peaceful oa-

The famed Eagle Gate Monument and the 28-story LDS Church office building.

sis in the heart of the city, an oasis that attracts some 4 million visitors annually, making the secluded city block the top tourist attraction in the state.

Anchored by the majestic multispired Salt Lake Temple, which required 40 years to build during the nineteenth century, Temple Square features continuous, free guided tours of the gardens, the architecturally elegant Assembly Hall, two fascinating visitors' centers, and the domed Tabernacle, home of the famed Mormon Tabernacle Choir and the great 12,000-pipe Tabernacle Organ.

Temple Square's popularity is not limited to the peak vacation months of summer. In fact, the Christmas season is one of the busiest times of the year. That's when more than 300,000 tiny pastel lights sparkle in the trees and shrubs and along the walks and buildings as part of a Christmas-lights tradition dating back to 1965.

The influence of the city's founding pioneers is also apparent outside of Temple

Square. Directly across the street to the west of Temple Square are the Museum of Church History and Art and the Family History Library. Both facilities are open to the public without charge. In the library thousands of people daily pursue the fascinating challenge of seeking out their "roots" in the world's largest genealogy repository, while the museum offers a fascinating menu of art exhibits and lectures.

The neighboring block to the east of Temple Square is home to the towering Church Office Building; the stately Church Administration Building; the attractive women's auxiliary building, which houses general offices of the women's and children's organizations of the church; and the Beehive House and Lion House, restored residences of early colonizer/church leader Brigham Young. As is its neighbor to the west, this tree-laden block is punctuated with fountains, statuary, and finely manicured gardens.

Free guided tours are available in the

28-story Church Office Building, which features an observation deck providing a panoramic view of the valley.

The Beehive House and the Lion House have been restored to appear as they did when occupied by Brigham Young. The former was his primary residence and, together with his adjoining office, is open to free tours. The latter is now used for luncheons, wedding receptions, and other social occasions.

On the same block is the former Hotel Utah, an historic old structure being remodeled to provide church office space as well as public areas designed to welcome visitors to the city and make their stay in the community more impressive and more inviting.

The Church of Jesus Christ of Latter-day Saints is supportive of community efforts to preserve Salt Lake City as an attractive, wholesome, and economically healthy place to live, work, and raise families. Its contributions are evident in its emphasis on the teaching of moral values and its encouragement of strong families and good citizenship.

Attractive Latter-day Saints meeting-houses dot the valley, serving the needs of hundreds of neighborhood congregations. With its headquarters offices and its widespread education, and meeting-

ABOVE: Assembly Hall on Temple Square is the site of the weekly Temple Square Concert Series.

BELOW: Every day thousands of people research their genealogies at the Family History Library.

house construction programs, the church also is a leading area employer.

The community's convention center and cultural arts complex sit on property provided by the church, and a real estate arm of the church has built several attractive downtown office and apartment buildings and an indoor shopping mall, all adding to the economic vitality of the community.

Zions Cooperative Mercantile Institution (ZCMI), a major downtown department store, was founded by Brigham Young soon after the Latter-day Saints settled the valley and today includes a network of suburban stores as well as the downtown flagship, with its architecturally pleasing facade.

In a tradition started by the early pioneers, music, dance, and theater continue to play an important role in Salt Lake City life.

Perhaps the most famous of the city's cultural offerings is the world-renowned Salt Lake Mormon Tabernacle Choir. Barely a month after the arrival of the pioneers in 1847, a church choir was performing on the site where several years later the Tabernacle would be built. These were humble beginnings for what has evolved into the 325-member Tabernacle

ABOVE: *Jerold Ottley directs the 325-member Mormon Tabernacle Choir.*

TOP LEFT: *The Museum of Church History and Art is open to the public without charge.*

LEFT: *Formerly a vaudeville theater, the Promised Valley Playhouse was renovated by the Mormon Church.*

FACING PAGE, TOP: *During the Christmas season, Temple Square is lit with more than 300,000 tiny pastel lights that sparkle in the trees and shrubs and along the walkways.*

FACING PAGE, BOTTOM: *Thanks to recordings, concert tours, and radio and TV programs, the Mormon Tabernacle Choir has fans worldwide.*

Choir, an ensemble with fans worldwide, thanks to its recordings, concert tours, and radio and television programs. The choir's Sunday-morning radio broadcasts from Temple Square began in 1929, and the half-hour show is now the oldest continuously running network (CBS) program in American radio history. The 9:30 a.m. broadcast is free to the public, as are weekly Thursday-evening rehearsals.

Public recitals on the great Tabernacle Organ are presented at noon and 4 p.m. on weekdays and at 4 p.m. on weekends.

Another fine musical institution sponsored by the church is the Mormon Youth Symphony and Chorus. The 100-piece orchestra and 300-voice chorus are comprised of young Latter-day Saints ages 18 to 30. The ensemble has performed in concert both nationally and internationally, and is seen periodically on public television and other networks.

A continuing source of free but quality entertainment downtown is the weekly Temple Square Concert Series in the Assembly Hall. Every Friday and Saturday evening during the year, individuals and groups ranging from new and local talent to internationally acclaimed stars perform for visitors. During December the weekly series turns into nightly concerts of Christmas music featuring community and school choirs.

Then there is the Promised Valley Playhouse, a former vaudeville and movie theater that the church acquired and renovated to provide a center for training in the arts as well as performing. The playhouse specializes in family entertainment, and its musical and dramatic stage shows are free to visitors who obtain tickets on Temple Square. The playhouse also features annual events such as July's International Vocal Month—where artists from around the world train local vocalists—and June's piano classes, workshops, and lectures, which complement the respected Gina Bachauer Piano Competition held each year in Salt Lake City.

ST. MARK'S HOSPITAL

St. Mark's Hospital first opened its doors in 1872 in a rented adobe house with six beds and a staff of one. From those humble beginnings it has grown into a full-service health care center with 306 beds, more than 300 physicians, and 1,000 professional, technical, and service personnel.

Along the way St. Mark's established many firsts in Utah's medical history. In addition to being the first hospital in the territory, St. Mark's began the first nurses' training school, opened the first women's unit, brought the first X rays to Utah, and was the first hospital to boast an ambulance.

In the 1920s St. Mark's served as the first home for the Shriner's Hospital for Crippled Children. In 1948 a residency and intern program in surgery and orthopedics was started, and in the 1950s and 1960s St. Mark's became the first hospital in the Salt Lake area to provide an intensive care unit and general psychiatric care.

But the most important first in the hospital's history was its tradition of growing and changing to meet the needs of Utah's people. In 1973 St. Mark's made per-

Personal care at St. Mark's Hospital ensures the dignity of the patient.

haps its biggest change by moving from the building it had occupied for 100 years.

Its current location, at 1300 East and 3900 South, was chosen to better serve the growing population in the central and southern part of the valley. The new, ultramodern building reflects the progressive, yet warm, human attitude found throughout the hospital. Its award-winning design allows for separate, quieter patient units, more efficient surgical and support operations, and easy expansion capabilities. Even after more than a decade of operation, the building remains on the U.S. State Department's Tour for International Architects and Health Care Professionals.

Today St. Mark's Hospital continues its tradition of innovative growth. As a member of the Hospital Corporation of America (HCA), St. Mark's enlarges its commitment to the delivery of quality patient care. It offers a full spectrum of medical services, including intensive care units and 24-hour emergency and trauma care. St. Mark's has a modern helipad that receives trauma patients and others from a multiple-state area, adding to the more than 25,000 patients seen annually by the emergency department staff.

The hospital has developed a strong or-

thopedic department specializing in complete hip replacement and has received national attention for its microvascular surgery in limb reattachments. The clinical laboratory is equipped with the most up-to-date computerized diagnostic equipment, providing the patient with a quick, accurate, economical diagnosis. St. Mark's houses the sophisticated equipment necessary to support a broad range of procedures, including laser surgery, digital subtraction angiography, computerized axial tomography, percutaneous nephroscopy, and magnetic resonance imaging. St. Mark's acts as a community resource center for the many hospitals in Utah and surrounding states that lack such equipment and expertise.

New parents enjoy the beautifully decorated birthing rooms at the hospital. St. Mark's offers a popular, early-discharge program with a 12- to 24-hour labor, delivery, and recovery in a homelike environment. Traditional extended-stay deliveries are also available.

St. Mark's enjoys one of the finest, most highly skilled staffs in the area. More than 300 physicians specialize in areas ranging from neurology to cardiovascular surgery, to family and general practice. To ensure the highest-quality care available, physicians participate in continuing education, peer review, and quality assurance programs. All personnel know

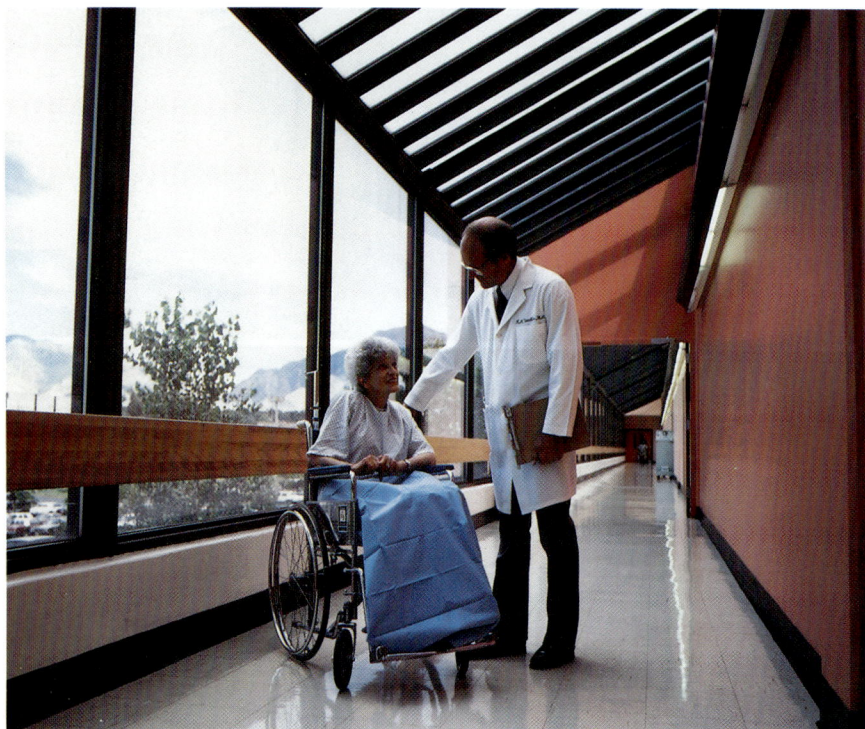

that patients come first. St. Mark's goal is to give the patient high-quality care, ensure human dignity, and create a comfortable environment where privacy is respected.

All hospital rooms at St. Mark's are private single care, with individual baths and beautiful views. They offer all the advantages of private rooms at costs competitive with multibed rooms at other hospitals in the area. At St. Mark's, patients enjoy privacy, flexible visiting hours, and a high degree of control over their hospital environment.

Cost-effective and convenient medical care is a high priority at St. Mark's. Efforts are also made to enable procedures to be done on an outpatient ambulatory basis whenever possible. A variety of convenient, economical services are available, and preadmission testing and ambulatory diagnostic services eliminate inpatient hospital expenses. Mini-stay surgery allows the patient to enter and leave the hospital the day of the operation without sacrificing safety. St. Mark's Ambulatory Intravenous Therapy Program eliminates prolonged hospitalization by allowing patients to continue therapy in their own homes, under hospital supervision. These programs have helped make St. Mark's one of the least expensive per-day hospitals in the Salt Lake metropolitan area.

The Community Education Department

at St. Mark's believes people who take responsibility for their own physical, mental, and social health are better prepared to stay healthy. A staff of physicians, nurses, dietitians, and social workers offers adult classes in weight control, smoking cessation, parenting, and physical fitness. It offers a great variety of health screenings that promote early detection and prevention of diseases such as glaucoma, high blood pressure, elevated cholesterol, and breast cancer. St. Mark's sponsors support groups that provide needed assistance with bereavement, and alcohol and gambling abuse. The Children's World of Wellness promotes a healthy physical and emotional life-style for children ages eight to 12 years, and a progressive childbirth education program completes a scope of classes that emphasize total family health care.

For more than 117 years St. Mark's Hospital has established an excellent reputation for leadership, care, and compassion. The institution looks forward to the challenge of the next century—of providing Utah and the Intermountain region with affordable, superior medical care.

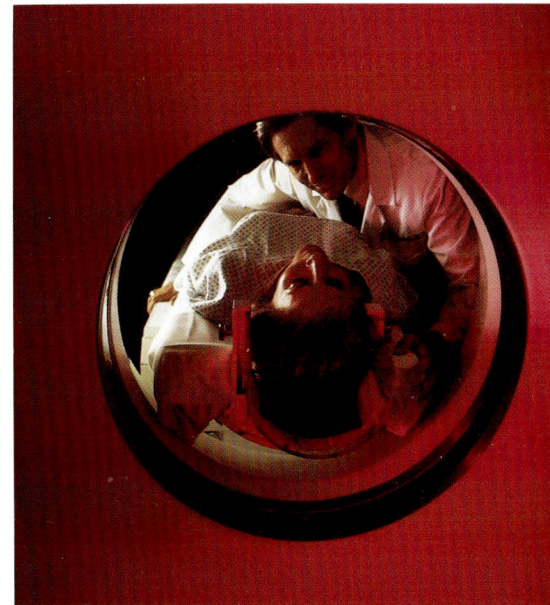

ABOVE: CAT and MRI scans are just two of the high-technology services available for patients at St. Mark's.

LEFT: St. Mark's, the first hospital in Utah, continues a tradition of progress.

BELOW: New parents enjoy St. Mark's comfortable birthing rooms.

CHAPTER THIRTEEN

THE MARKETPLACE

T he Salt Lake Valley's retail establishments, service industries, and products
are enjoyed by residents and visitors alike.

Solitude Ski Resort, 194

Crossroads Plaza Asso-
ciates, 195

The Snelgrove Ice Cream
Company, 196-197

Orleans Inn Hotel,
198-199

Photo by Mark E. Gibson

SOLITUDE SKI RESORT

For over a decade Solitude Ski Resort has offered outdoor enthusiasts the chance to "ski the freedom" on more than 1,000 acres of some of the most beautiful and exciting ski terrain in the world. Located 12 miles up Big Cottonwood Canyon in the Wasatch National Forest, the freedom of Solitude includes a wide variety of groomed trails and open bowls suited for all levels of skiers, from the first-time beginner to the aggressive expert.

Cross-country skiers at Solitude enjoy the same diversity as alpine enthusiasts. The Solitude Nordic Center offers miles of groomed ski trails through beautiful alpine forests, as well as guided backcountry trips. The Nordic Center is a natural rendezvous for ski tourers who wish to explore the abundant backcountry opportunities in the canyon, either on their own or on scheduled day and overnight guided tours.

Solitude's reputation for an absence of lift lines, friendly personnel, and reasonable prices has made it a favorite with local skiers and an increasing number of out-of-state visitors. It is the perfect family ski resort.

As Solitude enters its second decade,

its management includes Gary L. DeSeelhorst, president; Hal Louchheim, executive vice-president; Mike Goar, vice-president/mountain operations; and Mark Wilson, vice-president/village operations. This management team is undertaking an extensive program of replacing and upgrading its facilities with the objective of providing an improved recreational experience for all of its guests. "We're renovating Solitude to ensure that it continues to provide a quality and affordable ski experience well into the future," says DeSeelhorst. "Our master plan complements both the environment of the canyon and our desire to provide a little something for everyone."

Achieving better use of the mountain terrain required redesigning of the trail network, the creation of an exclusive first-time beginner area, relocation of two ski lifts, and the addition of a high-speed detachable quad chair lift. "This detachable quad chair lift, Utah's first, is a skier's dream," says Goar. "It delivers skiers back to the top of the mountain at twice the speed of conventional lifts, and plays a key role in maintaining Solitude's reputation for a lack of lift lines."

The second phase of Solitude's master plan calls for the complete replacement of all existing base facilities. "We will create a European-style, pedestrian-oriented village at Solitude," says Wilson. "It will include a variey of shops, restau-

ABOVE: Nestled high in the Wasatch Mountains of Utah, Solitude Ski Resort has a base elevation of 8,200 feet, ensuring ample quantities of Utah's famous snow.

LEFT: The stunning views and abundant powder skiing in Solitude's Honeycomb Canyon add an exciting dimension to any skier's day.

rants, lounges, and services for the day skier. There will also be limited overnight accommodations."

"Our goal is to create a year-round focal point for visitors in Big Cottonwood Canyon," adds Wilson. "The ambience of the village will be very friendly and unpretentious, in a style that enhances its beautiful alpine setting."

The village will serve as an attractive destination for canyon visitors and residents during the non-skiing season. Summer recreational activities include picnicking, hiking, fishing, guided nature hikes and lectures, children's day camps, and just relaxing in the cool mountain environment.

Solitude's improvements and its commitment to friendly customer service enable visitors to enjoy an experience comparable to any world-class ski area at incredible local prices. "We invite you to discover the freedom of Solitude," says DeSeelhorst. "It's the best-kept secret in Utah."

CROSSROADS PLAZA ASSOCIATES

Atriums, courtyards, glass-encased elevators, and a skylight that tops four levels of fine retail stores exemplify the exciting shopping environment at Crossroads Plaza. Strategically located just minutes from the airport, hotels, and points of interest, Crossroads Plaza is in the heart of Salt Lake City. Interconnecting four major city blocks, Crossroads Plaza is only steps away from Temple Square, the Salt Palace Convention Center and Professional Arena, the Salt Lake Arts Center, and Symphony Hall.

Weekday or weekend, day or night, Crossroads Plaza attracts shoppers, working people, visitors, conventioneers, sports fans, and concertgoers. Crossroads Plaza is in the center of a city filled with excitement, entertainment, and people.

The plaza offers a variety of stores, including Nordstrom, Weinstock's, and 145 specialty shops, services, and restaurants. The Richards Street level features an international food court, specialty retailers, and a three-screen cinema. The top of the plaza houses a sports fitness complex, complete with an indoor track, basketball and racquetball courts, nautilus, and rooftop tennis. Patrons enjoy eight levels of convenient, covered parking, free with validation.

Crossroads Plaza opened in 1980 as part of a mixed-use project in the thriving north end of the city that adjoins the 20-story Key Bank Tower and the 515-room Salt Lake Marriott Hotel. The facade of the Amusson Building, one of the oldest and most architecturally unique structures in Salt Lake City, was restored and now serves as the Main Street entrance to Key Bank. Crossroads Plaza quickly established itself as a major retail center in the Intermountain Region and is "Simply the Best" in fashion, food, and fun downtown.

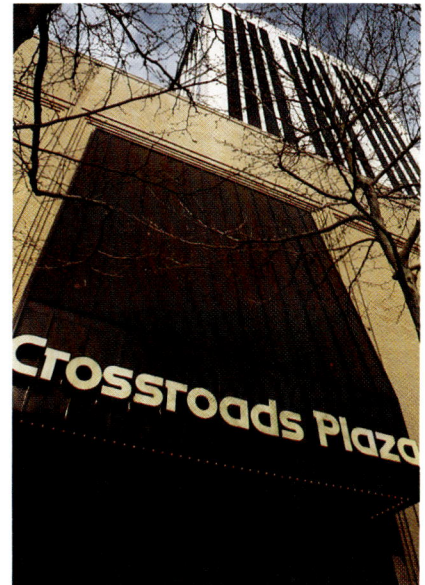

TOP: Crossroads Plaza offers a variety of well-known stores, including Nordstrom, Weinstock's, 145 specialty shops and services, and a recently renovated international food court.

ABOVE: Located in downtown Salt Lake City, Crossroads Plaza features "Simply the Best" in fashion, food, and fun.

LEFT: Crossroads Plaza adjoins the 20-story Key Bank Tower and the 516-room Salt Lake Marriott Hotel.

THE SNELGROVE ICE CREAM COMPANY

Utahans love ice cream! In fact, they eat more of it than anyone else, anywhere.

That is not surprising, considering the fact that Salt Lake City is the home of The Snelgrove Ice Cream Company.

Snelgrove's unparalleled ice cream—acclaimed in national magazines, specially ordered by presidents of the United States, recognized by the Smithsonian Institution, and considered Utah's most delicious tourist attraction—has been manufactured by the Snelgrove family for more than 60 years.

It is a tradition of taste that began in 1929, in the midst of the Great Depression. Charles R. Snelgrove; his wife, Fidella; and their sons, C. Laird and J. Barr, started an ice cream business with a good name and $5,000 in borrowed funds. The company's superior product and customer-oriented attitude quickly gained a devoted following. C. Laird Snelgrove took the company's helm in 1936.

Today, after six decades of growth, the third generation of Snelgroves is helping Laird and his wife, Edna, build upon the family tradition of quality, service, and taste. Six ice cream parlors sell Snelgrove ice cream, specialty ice cream desserts, and gourmet, hand-dipped chocolates. And Snelgrove ice cream is available in grocery stores throughout the West.

The recipe for Snelgrove's ice cream has been a well-kept secret for more than a half-century. But any ice cream connoisseur can quickly identify the rich, creamy desserts that are made in the Snelgrove plant behind the company's main store at 850 East 2100 South in Salt Lake City. The fresh taste and creamy texture is delightfully distinctive.

"Give the customer the best possible quality and service equal to that quality," explains Laird. "We've been successful because we make a superior product—with painstaking care, one batch at a time." Through the years the Snelgrove family has stubbornly resisted the industry's trend to lower price by lowering quality.

Snelgrove's quality has been maintained not only by using the freshest dairy products and best flavoring ingredients available, but also by continually reinvesting company profits into the new plant and equipment. Charles R. may

ABOVE: Laird and Edna Snelgrove celebrate 60 years of making America's finest ice cream.

RIGHT: A Salt Lake City landmark, Snelgrove's 15-foot double cone has whetted the appetites of kids and grown-ups alike for more than 50 years.

not recognize Snelgrove's modern manufacturing facility if he were alive today, but he would see one thing very clearly: The company he founded remains loyal today to the high production standards he established in 1929.

The insistence on quality has paid dividends—in 1988 the company achieved record sales of almost one million gallons of frozen desserts sold throughout the West. That is approximately 20 million ice cream cones a year. This statistic makes the Snelgrove family very proud. "People can't help but be happy when they're eating good ice cream," claims Laird. "I'd like to think we're responsible for a lot of happiness."

Take-out orders are always welcome at Snelgrove's, especially those from the White House. In 1935 the company sent several dozen gallons of chocolate and vanilla ice cream, packed in dry ice, via biplane to Warm Springs, Georgia. An ailing President Franklin D. Roosevelt was spending Thanksgiving in a Warm Springs polio hospital and wanted to share a special Thanksgiving dessert, Snelgrove ice cream, with the other patients. In 1988 President Ronald Reagan requested a batch of jelly-bean ice cream. Always aiming to please, Snelgrove's mixed up a special batch and sent it off to the White House.

But Snelgrove's fame in Washington extends well beyond the White House. Utah is known as the Beehive State, and a pewter mold used by Snelgrove to create made-to-order ice cream beehives is on display in the Utah exhibit at the Smithsonian Institution. Ice cream beehives have been served at National Press Club dinners, and have been frequently used by members of Utah's congressional

RIGHT: Soon to be shipped to ice cream lovers throughout the Intermountain region, Snelgrove all-natural ice cream passes through the final stage of production.

BELOW: Blending old-fashioned quality with state-of-the-art freezing equipment, production manager Phillip George tests ice cream for texture.

delegation to showcase their state's products.

Utah has been generous to Snelgrove, and Laird has expressed his gratitude and appreciation through strong support of local charities and sponsorship of community programs. Laird is a past president of the International Utah Visitors Council, where he and Edna host foreign dignitaries from the world of politics and entertainment. Laird has also served as

the director of the National Council of International Visitors and is an escort officer for the State Department. Distinguished guests from every continent have been treated to the sights and sounds of Salt Lake City, and to the incomparable taste of Snelgrove ice cream.

As Snelgrove Ice Cream Company celebrates its 60th anniversary in 1989, it is grateful to the many family members and employees who over the decades have made such generous contributions to the success of the company. Today the organization's co-founder, C. Laird Snelgrove, can proudly watch his company produce record amounts of his famous ice cream, gourmet desserts, and fine chocolates. In an intensely competitive industry, Snelgrove remains an industry leader with a commitment to bring the highest quality ice cream, confectionaries, and service to a growing number of valued customers.

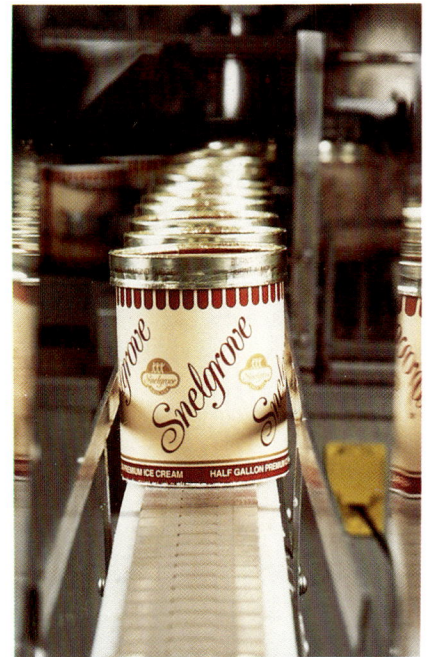

ORLEANS INN HOTEL

Located a stone's throw from the bustle of downtown Salt Lake City, the Orleans Inn Hotel is a quiet oasis of southern hospitality and New Orleans charm. The care and attention to detail at the family-owned inn have attracted a large contingent of loyal repeat customers. "We take weary businesspeople and give them a smiling face," says Mary Barney, office manager and family member. "Residents often comment on our friendly service and staff."

"Orleans Inn fills the gap between a hotel room and a furnished apartment," says owner Jerome Mooney. "We cater to the resident who stays a little longer than usual, people in town on extended business trips or for a long ski vacation. The average stay here is two to six weeks."

The Orleans Inn offers tourists special weeklong ski packages that include an evening social, travel arrangements to and from the ski shuttles, and a heated outdoor pool, called "our giant hot tub" by the hotel staff.

The hotel has been host to such distinguished guests as sculptor Karl Momen, who spent two years living at the Orleans Inn while he created his controversial *Tree of Life* in Utah's west desert; also, 20 engineers from Red China recently spent several weeks at the inn while they completed training at a

There is no such thing as a "room" at the Orleans Inn. One- or two-bedroom suites feature a living room as well as a full kitchen.

nearby computer facility.

The Orleans Inn serves about 60 large corporations on a continuing basis. Companies such as Chevron Oil, Delta Air Lines, and Mervyn's Stores regularly house their visiting executives at the inn.

There is no such thing as a "room" at the Orleans Inn. One- or two-bedroom suites feature a living room, as well as a kitchen with a standard-size stove and refrigerator. "Our kitchens are fully equipped," says Barney. "The traveler needs just a few personal items and can leave the kitchen sink at home."

Mooney conceptualized the Orleans Inn on a Mardi Gras vacation in New Orleans. He came home to Salt Lake City, found a quiet block within walking distance of downtown's best dining and entertainment facilities, bought the land, and made his dream a reality. "I designed

The backlit signed Tiffany window is the signature piece that marks the entry of the Orleans Inn.

the central courtyard and pool after the beautiful hotel where I stayed in New Orleans. I also borrowed the idea of every room being decorated differently. And I love the look of our wrought-iron balconies. They truly lend a southern accent to the place."

Mooney is especially proud of the exquisite stained-glass window that marks the entryway to the Orleans Inn. He found the signed Tiffany antique in Southern California and knew immediately that it would be perfect as his hotel's centerpiece. "The window is backlit at night and is a real standout. We consider it our signature piece."

Continental breakfast is included with all rooms. On winter mornings guests en-

brunch party as Pete's is in Salt Lake City.

"We're not big, we're not ultra deluxe," says Jerome Mooney, "but we boast about our southern hospitality. We're the friendliest hotel in town. We're successful because we're first class. And most of all, we're homey."

LEFT: Everything a corporate traveler needs can be found in an executive suite at the Orleans Inn.

BELOW: Guests may dive into a continental breakfast at poolside or just bask in the New Orleans charm.

joy the conservatory feel of the glassed-in patio where the morning meal is served. In warm weather breakfast is served poolside in the center courtyard.

If guests prefer, they may eat out or stay "inn." The Orleans Inn houses an exciting full-service restaurant called Pete's, unique in Salt Lake City for both its style and its menu. Pete's casual and relaxed atmosphere underscores the professional, attentive service and beautifully prepared food that has made the restaurant a favorite with locals and tourists alike.

Guests can enjoy Pete's menu, inspired by great recipes from around the world, for breakfast, lunch, dinner, late-night fare, and weekend brunch in any of the restaurant's several areas. For guests too exhausted from a hard day on the slopes to make their way to the dining room, Pete's staff is more than happy to provide room service—quickly and with a smile.

The Boom-Boom Room features local jazz performers and invites patrons to dance the night away. The glass-enclosed Copacabana Room offers a bright, cheerful patio environment, and al fresco dining is available poolside in the courtyard from early spring through autumn.

For the sophisticated shark, billiards are available at Pete's Pool, furnished exclusively with eight Golden West Tables by Quality Billiards, the Intermountain West's only billiard manufacturer.

Guests shouldn't go home without the recipe for Pete's World Famous French Toast—sure to be as big of a hit at any

As Salt Lake Valley races toward a new horizon in the twenty-first century, the challenge will be to preserve its natural beauty and open spaces, while accommodating inevitable population growth and commercial expansion. Photo by Stephen R. Smith

PATRONS

The following individuals, companies, and organizations have made a valuable commitment to the quality of this publication. Windsor Publications and the Salt Lake Area Chamber of Commerce gratefully acknowledge their participation in *The Salt Lake Valley: Crossroads of the West.*

American Savings and Loan Association

Arthur Andersen & Co.*

Babcock Pace + Associates, Architects*

Bonneville International Corporation
Broadcast House • KSL-AM • KSL-TV • Video West • Bonneville Media Communications*

Church of Jesus Christ of Latter-day Saints*

City Centre/A Price-Prowswood Development*

Clark-Leaming Corporation

Crossroads Plaza Associates*

Culp Construction Company

Cytozyme*

Delta Air Lines*

Deseret Medical, Inc., Becton Dickinson*

Deseret News*

Eastman Christensen*

Geneva Steel of Utah*

Growers Market Company

Hercules Aerospace*

Huntsman Chemical Corporation*

IBM*

Intermountain Power Agency*

Jetway Systems*

Johnson & Higgins*

Jones, Waldo, Holbrook & McDonough, P.C.*

KUTR-AM Radio*

Mansell & Associates

Mark Steel Corporation*

MCI Telecommunications Corp.

Mineral Mine*

Morton International*

Murdock Travel Management*

Nabisco Brands, Inc.

Natter Manufacturing Company/Fairchild Industries*

Nickerson Company, Inc.

Orleans Inn Hotel*

Peat Marwick Main & Co.*

Questar Corporation*

St. Mark's Hospital*

Salt Lake Convention & Visitors Bureau

Salt Lake Investment Company

Savage Industries, Inc.*

William Selvage

The Snelgrove Ice Cream Company*

Snow, Christensen & Martineau*

Solitude Ski Resort*

Southmark Equities Corporation

Stabro Laboratories, Inc.*

Stat Express Inc.

Synergetics Realty Company, Inc.

Technical Devices Company

Terra Tek, Inc.

Thiokol Corporation*

3M Health Information Systems*

Utah Economic Development Corporation*

Utah Power & Light Company*

Van Cott, Bagley, Cornwall & McCarthy*

Varian Eimac*

Western Rehabilitation Institute

Western Savings of Arizona

*Salt Lake Valley's Enterprises in *The Salt Lake Valley: Crossroads of the West.* The stories of these companies and organizations appear in Part Two, beginning on page 123.

BIBLIOGRAPHY

CHAPTER ONE

Alexander, Thomas, and James B. Allen. *Mormons and Gentiles.* Boulder, Colorado: Pruett Publishing Company, 1985.

Del Porto, Brett. "Utah Population Boom Likely to Quiet Down." *Deseret News,* June 2, 1987.

Gutman, John. "Dr. Willem J. Kolff, Modern Pioneer." *Salt Lake Tribune,* January 19, 1986.

Harmer, Mabel. *The Story of the Mormon Pioneers.* Salt Lake City: Deseret News Press, 1943.

Korologos, Michael. "Utah: New Energy, New Directions." *Western's World,* August 1985.

"A Look at Utah's Labor Market in 1986." *Utah Annual Report 1986.* September 1987.

"Only Utah, New Mexico Show Jobs Expansion in Region." *Deseret News,* August 20, 1987.

"Putting Salt Palace in Downtown S.L. Was the Best Decision, Planner Says." *Salt Lake Tribune,* November 14, 1987.

Rolando, Joe. "High Hopes for Utah's High Tech." *Salt Lake Tribune,* November 18, 1987.

Smart, William B. "Utah's Future: Coalition of Concerned Citizens Seeks to Break Down Many Societal Barriers." *Deseret News,* November 24, 1987.

"South and West See Most of the Nation's Population Growth." *Deseret News,* January 1, 1987.

Twain, Mark. *Roughing It.* New York: Signet, The New American Library Inc., 1872.

Walters, Conrad. "DePaulis Unveils Blueprint for Building a Future." *Salt Lake Tribune,* June 24, 1987.

Woody, Robert H. "Jordanelle Reservoir Under Way." *Salt Lake Tribune,* November 1, 1987.

CHAPTER TWO

"Area Home Sales Rose 6% in 1986, and Average Price Increased 1.7%." *Deseret News,* January 19, 1987.

Korologos, Michael. "Salt Lake City 1984: Thinking Big, Doing Big." *Western's World,* August 1984.

Palmer, Douglas D. "Rabbi Praises S.L. Emphasis on Faith, Education." *Deseret News,* November 24, 1987.

Sheehy, Sandy. "Salt Lake City." *Town and Country,* August 1983.

"S.L. Police Department Commended as one of the Best in the Country." *Deseret News,* November 26, 1986.

U.S. Department of Justice, Federal Bureau of Investigation. *Uniform Crime Reports, Crime in the United States.* July 1986.

"Utah Ranks Among the Five Most Desirable Regions." *Deseret News,* January 28, 1985.

"Utah's Per Capita Income Will Remain Low Despite Predicted Increase in Jobs." *Deseret News,* November 24, 1987.

Wilson, Ted. *Utah's Wasatch Front.* Salt Lake City: Utah Geographic Series, November 1987.

Wood, James A. "The Utah Housing Market." *Utah Economic and Business Review,* March 1987.

CHAPTER THREE

Boren, Ray, and Pam Wade. "Lassoing Tourists." *Deseret News,* October 18, 1987.

Boulton, Guy. "Travelers Hopes to Restore Triad's Luster." *Salt Lake Tribune,* May 7, 1987.

————. "Triad Files Under Chapter 11." *Salt Lake Tribune,* January 29, 1987.

Brown, Matthew. "Bangerter Urges Investors to Take a Fresh Look at Utah." *Deseret News,* January 21, 1988.

"DePaulis Calls for Plan to Lift Economy." *Salt Lake Tribune,* January 20, 1988.

"11 Firms Selected for Venture Capital Session Jan. 28." *Deseret News,* January 2, 1988.

Flanagan, Barbara. "Portland the Perfect." *Metropolitan Home,* January 1988.

Funk, Marianne. "Utah's Bad-mouthers No Help To Economy." *Deseret News,* January 21, 1988.

"Hercules Hopes to Propel into Forefront of Missile and Shuttle Motor Production." *Deseret News,* March 15, 1987.

Hyatt, Joshua. "Inc.'s Annual Report on the States." *Inc.,* October 1987.

"Inc.'s Annual Report on the States." *Inc.,* October 1985.

Knudson, Max B. "Convention Mecca: S.L. Convention and Visitors Bureau Doing A-1 Job of Proving to Travelers That Utah is Great Place to Visit–Summer or Winter." *Deseret News,* November 9, 1986.

————. "S.L. Economic Potential Nets a Cinderella Ranking." *Deseret News,* December 5, 1984.

Lamb, David. "Salt Lake: A Worship of Order." *Los Angeles Times,* February 3, 1988.

McMullin, Eric. "Convention Bureau Reorganized to Go After More Bucks." *Salt Lake Tribune,* January 28, 1988.

————. "Utah Needs an Image, Says Tourism Survey." *Salt Lake Tribune,* October 17, 1987.

Naisbitt, John. *Trend Letter.* Washington, D.C.: Global Network, January 7, 1988.

Oberbeck, Steven. "Salt Lake Economy Banks A Lot on Fed Branch." *Salt Lake Tribune,* August 30, 1987.

Pusey, Roger. "Report on Year 2,000 Stresses Need for Skilled Work Force." *Deseret News,* February 12, 1988.

————. "S.L. Convention and Visitors Bureau Predicts an Even Better Year in 1988." *Deseret News,* January 28, 1988.

————. "Utah Must Halt 'Bad Signals,' Farley Says." *Deseret News,* February 11, 1988.

Rolando, Joe. "Geneva Head Urges New Climate for Business Growth." *Salt Lake Tribune,* January 26, 1988.

————. "High Hopes for Utah's High Tech." *Salt Lake Tribune,* November 18, 1987.

————. "S.L. Expects to Land a Developer for Block 57 in '88, Says Mayor." *Salt Lake Tribune,* February 17, 1988.

Spangler, Jerry. "Voyager's Flight is Expected to Conquer a Final Frontier." *Deseret News,* December 5, 1984.

"Westin Hotel Utah, an S.L. Landmark, Will Close in August." *Deseret News,* March 13, 1987.

Woody, Robert H. "Douglas Plant is Symbol of Utah's Quest to Become a Space Center." *Salt Lake Tribune,* January 17, 1988.

————. "Hercules Dedicates Huge Rocket Motor Plant." *Salt Lake Tribune,* March 11, 1987.

————. "Utah May See a Little Growth, But Future Depends on U.S." *Salt Lake Tribune,* January 8, 1988.

Woolf, Jim. "Utah's Hotel of the Century to Close." *Salt Lake Tribune,* March 13, 1987.

CHAPTER FOUR

Atchison, Sandra D. "Meet the Campus Capitalists of Bionic Valley." *BusinessWeek,* May 5, 1986.

"Deseret Medical Chief Aims to Make Community Aware of Firm." *Salt Lake Tribune,* May 3, 1987.

"High Tech America: Where it Is, Where It's Going." *Grubb & Ellis Market Trends,* August 1987.

Jacobsen, JoAnn. "Park's Official Urges Conference Center." *Salt Lake Tribune,* July 16, 1974.

Love, Perrin. "U Park Rated a Success." *Deseret News,* March 24, 1977.

Palmer, Anne. "Grant Pumps Millions Into U's Artificial Heart." *Salt Lake Tribune,* January 7, 1988.

Spangler, Jerry. "Big Bucks Called the Key to Making Utah Center of High-Tech Development." *Deseret News,* September 23, 1986.

"Terra Tek Chief Knows Value of Campus-Business Relationship." *Salt Lake Tribune,* September 21, 1986.

CHAPTER FIVE

Bureau of Business and Economic Research. "Economic Impact of the Utah Transit Authority." January-February 1988.

"Curing I-15itis. Light-Rail System and More Freeway Lanes Won't Come Cheap or be Panaceas." *Deseret News,* March 13, 1988.

Davidson, Lee. "Traffic at S.L. Airport Expected to Double in Next 13 Years." *Deseret News,* September 10, 1987.

"Delta to Begin Construction of Hangar for Service, Use." *Salt Lake Tribune,* April 11, 1988.

"Delta-Western Merger Takes Off After Turbulent Delay." *Salt Lake Tribune,* April 2, 1987.

Evensen, Jay. "Rapid Transit Rail System Urged as Way to Boost Sales and Revitalize Downtown." *Deseret News,* November 3, 1987.

Fagg, Ellen. "Growth is Key Word at S.L. Airport." *Deseret News,* March 23, 1987.

————. "S.L. Airport Among the Nation's Most Punctual." *Deseret News,* December 17, 1987.

Fenton, Jack. "S.L. Alliance Backing Light Rail." *Salt Lake Tribune,* April 4, 1988.

Fitzpatrick, Tim. "Airport's 20-Year Master Plan Lands Look of 21st Century." *Salt Lake Tribune,* May 8, 1988.

Gorrell, Mike. "FAA Seeks Ways to Upgrade Capacity and Efficiency of Salt Lake Airport." *Salt Lake Tribune,* November 30, 1987.

"Light Rail Could Serve S.L. Area by '94 at Earliest, UTA Board Told." *Deseret News,* January 21, 1988.

Pingree, John. "Salt Lake Has Winning Mass Transit." *Salt Lake Tribune,* October 12, 1986.

"Put Your Fears to Rest About Delta's Impact in Salt Lake, Spokesman Urges." *Deseret News,* September 11, 1986.

"Rail Service Would Put Commuters on the Right

Track, Train Buff Says." *Deseret News*, April 25, 1986.

Salt Lake Airport Authority. *Centerline*, February 1988.

"S.L. Airport to Build New Delta Hangar." *Salt Lake Tribune*, April 12, 1988.

Wagner, Rodd G. "S.L.-Area Traffic Entering 'Crisis' As I-15 Overflows." *Salt Lake Tribune*, September 8, 1985.

CHAPTER SIX

Bureau of Business and Economic Research. "The Economic Impact of the University of Utah's Health Sciences Center." November 1986.

Funk, Marianne. "The Private Choice." *Deseret News*, December 14, 1986.

"Pupils Do Well in Utah Despite 'Underfunding.'" *Deseret News*, September 17, 1986.

Salt Lake Area Chamber of Commerce. "Report on Public Education in Utah." November 1987.

Scarlet, Peter. "Data Show Tough School Standards Dictate Class Choices." *Salt Lake Tribune*, November 10, 1987.

————. "Utah's Students Still No. 1 in Advanced Placement." *Salt Lake Tribune*, December 16, 1987.

Tracy, Dawn. "ACT Scores Drop in Utah; Are Big Classes to Blame?" *Salt Lake Tribune*, September 23, 1987.

————. "Utahns' ACT Scores Take a Jump, Except in Math." *Salt Lake Tribune*, September 25, 1986.

Utah Foundation. "Educational Expenditures in Utah—1986." *Research Briefs*, March 1987.

"Utah is No. 1 in Literacy—But Aims to do Still Better." *Deseret News*, April 24, 1986.

Utah State Board of Education. *Annual Report of the State Superintendent of Public Instruction, 1986-1987.*

Van Leer, Twila, and JoAnn Jacobsen-Wells. "IHC." *Deseret News*, April 5, 1987.

CHAPTER SEVEN

Free, Cathy. "Koch Strolls S.L., Calls it 'Cleanest City This Side of Heaven.'" *Salt Lake Tribune*, June 14, 1988.

Jordan, Fred. "Arts Districts Can Paint Downtowns the Color of Money." *Governing*, July 1988.

Krist, Gary. "Salt Lake: City of the Saints." *National Geographic Traveler*, Winter 1986/87.

Lincoln, Ivan. "Theater at the U. Marks its 25th Anniversary." *Deseret News*, September 25, 1987.

Rolly, Paul. "S.L. Shelter Project Will be Focus of Tour by 175 Visiting Mayors." *Salt Lake Tribune*, June 14, 1988.

"R/UDAT: The Choices are Before Us." *The Event*, July 1-15, 1988.

Weeks, Russell. "R/UDAT: City Must Keep Historic Character." *Salt Lake Tribune*, June 19, 1988.

CHAPTER EIGHT

Fenton, Jack. "Supertunnel Proposed for Wasatch Front." *Salt Lake Tribune*, August 25, 1987.

"Hearing: Olympics Not Only for Local Canyons." *Salt Lake Tribune*, August 2, 1988

Rolando, Joe. "Chamber Supports Solitude Construction, But 3 Citizens' Groups Have Appealed." *Salt Lake Tribune*, February 10, 1988.

INDEX

Salt Lake Valley's Enterprises Index

Arthur Andersen & Co., 168

Babcock Pace + Associates, Architects, 181

Bonneville International Corporation Broadcast House • KSL-AM • KSL-TV • Video West • Bonneville Media Communications, 134-135

Church of Jesus Christ of Latter-day Saints, 186-189

City Centre/A Price-Prowswood Development, 182-183

Crossroads Plaza Associates, 195

Cytozyme, 146

Delta Air Lines, 126-127

Deseret Medical, Inc., Becton Dickinson, 162

Deseret News, 133

Eastman Christensen, 176

Geneva Steel of Utah, 148-150

Hercules Aerospace, 142-143

Huntsman Chemical Corporation, 158-159

IBM, 172-173

Intermountain Power Agency, 130-131

Jetway Systems, 156-157

Johnson & Higgins, 169

Jones, Waldo, Holbrook & McDonough, P.C., 177

KUTR-AM Radio, 136-137

Mark Steel Corporation, 152-153

Mineral Mine, 140-141

Morton International, 144

Murdock Travel Management, 178-179

Natter Manufacturing Company/Fairchild Industries, 151

Orleans Inn Hotel, 198-199

Peat Marwick Main & Co., 180

Questar Corporation, 128-129

St. Mark's Hospital, 190-191

Salt Lake Area Chamber of Commerce, 166

Savage Industries, Inc., 154-155

Snelgrove Ice Cream Company, The, 196-197

Snow, Christensen & Martineau, 170-171

Solitude Ski Resort, 194

Stabro Laboratories, Inc., 160-161

Thiokol Corporation, 145

3M Health Information Systems, 174-175

Utah Economic Development Corporation, 167

Utah Power & Light Company, 132

Van Cott, Bagley, Cornwall & McCarthy, 176

Varian Eimac, 147

General Index

Italicized numbers indicate illustrations.

A

ABC, 72

Advanced Combustion Engineering Center, 57-58, *58*

Advanced placement courses, 77

Aerospace industry, 44

"Aerospace Target Industry Study," 44

Aircraft manufacturing, 65

Air Force bases, 20

Airport Master Plan, 65

Air quality, 25

Allen, R.W., 65

Alta, *109,* 114, 116

American Chamber of Commerce Researchers Association, 33, 34

American Towers, 34

Antelope Island, *30-31*

Apartments, 34

Arches National Park, 118

Area Vocational Centers, 79

Art galleries, 103

Artificial heart research, 22, 23, 57, 60, *60,* 61

Artificial organ research, 23, 57, 60, 61

Arts, 97-103; children's arts programs, 103; dance, 98, 103; movie industry, 101; music programs, 98, 100-101, 103; studio space, 101-102, *102;* theater, 98-100

Artspace, 102, *102*

Atlantis, 59

Autumn Aloft, 105, *105*

Avenues district, 34

B

Bacchus, T.W., 43

Bacchus Works, 43-44

Bailey, Bob, 115

Ball, Fred S., 46, 53, 56, 89

Ballet West, 79, 92-93, 98

Bangerter, Norman, 24-25, 49, 57, 61, 97, 117-118

Banking, 48

Barker, Bart, 41

Baseball, 103

Basketball, 103, *103*

Bass, Richard, 105, 114

Bates, George E., 107

Bayle, Deborah S., 79

Beneficial Life Tower, 47

Bennett, William, 77

Bertoch, Brad, 49

Bingham Mine, 19, *19,* 45, *45*

Biomedical Centers, 57, *57*

Biomedics, 22, 23, 57, 60, 61

Biotechnology Centers, 58

Birthrate, 74

Block 57, 47, 48

Bonneville International, 72

Bountiful City, 70

Boyer Company, 48

BP Minerals, 19, 22, 43, 121

Brande, Jim, 120-121

Brickyard Plaza, 36

Bridger, Jim, 15

Brigham Young University, 58, *77,* 84

Brighton, 114

Brophy, James J., 56-57, 88-89

Brown, Norman, 57

Brown Bag Concert Series, 103

Bryce Canyon National Park, 119

Business parks, 47-48

Business schools, 86

Bus system. *See* Public transit

C

Caldwell, Bruce, 92-93

Camp Floyd, 16

Camp of Israel, 15

Cannon, Joseph A., 40, 41

Canoeing, *78, 112*

Canyonlands National Park, 118, *121*

Canyons, 92, 110, *114-115,* 115, 116, *116*

Capital Reef National Park, 119, *119*

Capitol Theatre, 98-100, *99;* ghost of, 99-100

Career Ladder Program, 80

Catalyst, 73

Cathedral of St. Mary Magdalene, 92

Cathedral of the Madeleine, 36, 92, *94-95, 95,* 96

Catholics, 18, 92; schools, *82,* 83

CBS, 72

Center for Advanced Coal Technology, 58

Center for Controlled Chemical Delivery, 58, *58*

Center for Engineering Design, 57

Center for Materials and Advanced Manufacturing Technologies, 57

Center for Women's Health, *32*

Center of Biopolymers at Interfaces, 57

Centers of Excellence Program, 57-58

Central Pacific Railroad, 16, 68

Children's arts programs, 103

Children's Dance Theater, 103

Children's Museum of Utah, 103

Churches, 36, 92, *94-95,* 96

City and County Building, 18-19

"City Beautiful" campaign, 19

City Centre, 48
Clark, Barney, 22, 61
Clark Leaming Designs for Business, 38
Clearfield Naval Supply Depot, 69
Cleveland State University, study of Utah
    by, 28, 32
Cliff Lodge, 114
Cliff Spa, 105
Climate, 110
Climbing competition, 114
Cloud seeding, 25
Coal, 58
Coalition for Utah's Future, 26
Colmek Systems Engineering, 23
Colorado River, 118
Comes, John, 92
Communications, 58, 72-73
Communications Research Center, 48
Community Development Block Grant,
    102
Computer-assisted instruction, 80
Computer literacy program, 84
Computer technology, 58, 59
Condominium projects, 34
Construction, 46-48
Continental Bank and Trust, 48
Convention and Visitors Bureau, 50
Convention business, 50, 52, 53
Copper industry, 19, *19*, 21, 22, 43, 45,
    45, 62
Core Curriculum, 78-79
Cost of living, 33, 34
Cottonwood Hospital Medical Center, *12,
    89*
Cottonwood Mall, 36
Creekside Mall, 36
Crime rate, 37
Crossroads Mall, *46*
Crossroads Plaza, 36
Culture, 19, 97-103
Cushman, Trevor, 98
Custom Fit Training, 79

D
Dams, 25-26, *25*
Dance companies, 98
Dance programs, 103
Davis, Rick, 50, 52
Days of '47 Parade, 105, 107
Deadhorse Point State Park, 118
DeBries, William, 22
Deep Creeks, 62
Defense industry, 21, 43, 44, 45
Delta Air Lines, *64,* 65, *65, 66*
Denver Rio Grande Western Railroad, 68;
    depot of, 96

Department stores, 36, 49
DePaulis, Palmer, 26, 47, 53, 76, 109
Depression, the Great, *19,* 20
Depression (1890s), 18
Deseret Medical, 60
Deseret News, 12, 18, 73, 79
Desert, 119-120
Developers, 47-48
Devereaux House, *97*
Dimond, Margaret, 90
Downtown, 21, 46-48
Downtown Planning Association, 21
Draper City, 34
Dugway Proving Grounds, 30
Dynamite manufacturing, 43

E
Eagle Gate, 34
Eagle Gate Plaza and Office Tower, *40,
    46, 46,* 47
East Bench Corridor, 87
Eastern Airlines, 66
Eccles, George, 88
Eccles, Spencer F., 88
Eccles Foundation, 88
Eccles Institute of Human Genetics, 61,
    87
Economic development programs, 26, 41,
    53; cooperation between public and pri-
    vate sectors, 41
Edmunds Act, 18
Education, 12, 74, 76-81, 83-87; advanced
    placement courses, 77; budget,
    71-73; computer-assisted instruction,
    80; Core Curriculum, 78-79; enroll-
    ment, 80-81; extracurricular activi-
    ties, 79; fiscal problems, 24-25;
    funding, 74, 76, 81, 83; gifted pro-
    grams, 84; higher education, 58, *77,*
    84-87; high schools, 74; instructional
    television, 80; literacy rate, 74; preven-
    tion programs, 80; private schools,
    83; ranking in U.S., 77; school clo-
    sures, 81; school districts, 81; spe-
    cial education, 79; special programs,
    79-80; Sterling Scholar program, 79.
    See also University of Utah
Electricity, 70
Employment, 22, 41, 42, 60, 87, 89;
    growth predictions, 22, 43; women
    in work force, 42. *See also* Unemploy-
    ment
English, John, 72
Entrepreneurship, 38, 58, 79
Environmental programs, 120-121
Episcopalians, 18; schools, 83

Esch, Leigh Von de, 101
E-Systems Inc., 44
Ethnic festivals, 105-106
Evans, David C., 59
Evans & Sutherland, 40, 54, 56, 59, *59*
The Event, 73
Exportation, 68, 69

F
Family History Library of the Church of
    Jesus Christ of Latter-day Saints, 107
Farnsworth, Philo T., 72
Fashion Place Mall, 36
Federal Express, 65
Federal Procurement Office, 44-45
Federal Reserve Bank of San Francisco
    in Salt Lake, 48
Federal Urban Mass Transit Authority, 72
Fertility rate, 74
Festival of Trees, 106
Festivals, 105-106
Financial institutions, 48-49
First Interstate Bank, 48
First Presbyterian Church, 36
First Security Corporation, 48
First United Methodist Church, 36
Fish Springs National Wildlife Refuge, 62
Fletcher, James C., 54
Flextrans, 71
Floods of 1983, 108, 117
Flora and fauna, 62
Food processing, 69
Football, 104
Foothill Village, 36
Foreign Trade Zone, 68
Forum, 48
Freeport Center, 69
Freeport laws, 68
Freight carriers, 68-69

G
Genealogical library, *76*
Genetic research, 61, 87-88
Geneva Steel, 40-41, 43
Geography, 25, 28, 62
Georgalas, Dino, 42, 107
Globesat, 58
Golden Eagles, 103
Goldsmith, Stephen, 102
Government, 33, 40; contracts, 44-45, 60;
    regulation, 48-49
Government Affairs Council, 41-42
Governors Plaza, 34
Grand Canyon, 120
Grand Teton National Park, 120
Granite School District, 77, 81

Grant, Heber J., 21
Graphic computer systems, 59
Great Basin National Park, 120
Great Salt Lake, *27, 28, 30, 31,* 117-118
Great Salt Lake Yacht Club, *119*
Greek Festival, 105-106
Green, Sidney J., 60
Green River, *78,* 118
Grocery chains, 49
Growth: employment, 22; planning,
    26, 120; predictions, 24; problems,
    18, 19
Gustavson, Dean L., 46
Gymnastics, 86, 104

H
Health care, 18, 22, *22,* 23, *32, 74-75,*
    87-91; alternative systems, 90; employ-
    ment, 87, 89; genetic research,
    87-89; muscular dystrophy treatment,
    89; nursing homes, 89-90; psychiatric
    care, 90. *See also* Hospitals
Health Sciences Center, 22, *54-55,* 87, 89
"Healthy Utah," 90-91
Heavy industry, 42-43
Hercules, 23, 43-44
High schools, 74, *76,* 77-78, 81, *82,* 83
High technology, 57, 58, 59, 60; em-
    ployees, 60
Hiking, 116, 119-120
Hill Air Force Base, 20
Hillcrest High School, *76*
Hillside Plaza, 36
Historic preservation, 92, *94,* 96-97, *97,*
    99
Hockey, 103
Holy Cross Hospital, 18
Holy Trinity Greek Orthodox Church, 36
Homelessness, 108
Hospitals, 12, *12,* 18, *22, 32,* 60, 87-88,
    87, 89, *89,* 90, 91; admissions, 90; av-
    erage length of stay, 91. *See also*
    Health care
Hot-air balloon festival, 105, *105*
Hotels, 52, 53, *56,* 102
Housing, 33-34; shortages, 21; average
    price, 33
Hughes Medical Center, 87-88
Hundley, "Hot Rod," 103
Hunt, Duane G., 21

I
Immigration Visitors District, 86
Importation, 68-69
Inc. magazine's "Report on the States,"
    38, 42

Individual Education Program, 79
Industrial parks, *67,* 69
INERAID cochlear implant, 60
In-migration, 15-16, 21, 24
Instructional television, 80
Intermountain Health Care, Inc., 90
Intermountain Trauma Complex, 90
Interstate 15, *68-69,* 70, *70-71*
Interstate 80, *62-63,* 68
Interstate 215, *70*
Investment opportunities, 48-49
IOMED, 60

J
Jackson, Bill, 65
Jacobsen, Stephen C., 60
Japanese-American Citizen's League, 20
Jarvik-7 artificial heart, 57, 60, *60*
Jason Jr. (robot), 23
Jordanelle Dam, *25,* 25-26
Jordan School District, 81
Judge Memorial Catholic High School, *82,*
    83

K
Kahler Corporation of Minnesota, 56, 58
KBYU, 73
Kearns, Thomas, 17
Kennecott, 19, 21, 22, 43, 45, *45*
Key Bank, 48
Koch, Ed, 92
Kol Ami Synagogue, 36, 37
Kolander, Eugene, 83-84
Kolff, Willem, 23, 54, 60
KSL Radio, 72-73
KSL Television, 72, 79
KSTU, 73
KTVX, 72
KUED, 73
KUTV, 26, 73

L
La Caille at Quail Run, 104
Lady Utes gymnastics team, 86, 104
Lallapalooza, *84,* 103
LaManna, Lisa, *92-93*
Laser Institute, 57, *57*
Latter-day Saints. *See* Mormons
LDS Church, *27*
LDS Hospital, 90
Leaming, Merlene, 38, 92
Levitt, William H., 116
Liberty Park, *10-11, 23, 108*
Libraries, *76, 85,* 107
Lieske, Joel A., 28, 32
"Life Issues," 98

Life-style, 24, 28, 32, 37, 53, 92, 109,
    120-121
Light-rail system, 47, 70, 72
Lindsay, Richard, 26
Literacy rate, 74
Little Cottonwood Canyon, 92,
    114-115
Litton Guidance and Controls, 44
Living Traditions Festival, 106
Lodging. *See* Hotels
Lowell Elementary, *76*
Lutherans, 83

M
Magazines, 73
Magna, 23, 43, 44
Main Street, *17*
Malls, 36, *46,* 49
Manufacturing, 43-45, 60
Manufacturing Technology Centers, 57
Market Street Broiler, *22, 102*
Marriott Corporation, 42
Marriott Library, *85*
McDonnell Douglas Aircraft Company, 44,
    65
Media, 72-73
Megatrends, 38
Microelectronics Center, 58
Migration, 15-16, 21-22, 24
Mill Creek Canyon, *116*
Miller, Lou, 65, 67
Mining, 19, *19,* 21, 22, 43, 45, 62
Minor, Kent, 121
Missile manufacturing, 43-44
Morgan, Doug, 99, 100
Mormons, 12, 14, *14,* 15, *15,* 16, 17, 18,
    26, 38; arrival in Salt Lake, 15-16;
    art, 97-98; church headquarters, 47;
    during Depression, 20; evacuation of
    Salt Lake City (1857), 16; library,
    107; Mexican War, 15; percentage of
    population, 26; polygamy, 18; rela-
    tions with non-Mormons, 16, 20-21;
    trek to Utah, 14-15
Mormon Tabernacle Choir, 72-73, 97-98,
    98
Morris, June, 38, 109, 120
Morris Air Service, 38
Morris Travel, 38
Moss, James R., 78, 79, 83
Motion Control, 60
Mountain Fuel, 69
Mountain ranges, 28, 62
Mount Nobe Wilderness area, *7*
Mount Olympus, 35
Movie industry, 101

Murray City, 70
Murray City School District, 81
Muscular dystrophy treatment, 89
Museums, 86, 103
Music programs, 98, 100-101, 103

N
Naisbitt, John, 24, 38
National parks, 53, 118-120, *119, 121*
Native Plants, Inc., 54, 59-60
Natural gas, 69-70
NBC, 73
Neighbor Fair, 105, *107*
Nelson, David E., 77
Network, 73
New Deal, *19,* 20
Newspaper Agency Corporation, 73
Newspapers, 18, 73
New York Club, The, 102, *102*
Northwest Pipeline, 54
Nursing homes, 89-90

O
Odle, James F., 44-45
Office buildings, 46, 47-48
Office space, 46
Office of Technology Transfer, 56
Ogden, 17, *32-33*
Oktoberfest, 106
Oil, 21, *44,* 54
Olsen, Don B., 61
Olympic Gymnastics Trials, *103,* 104
Opera, 79
Oquirrh Mountains, 28, 62
Oswald, Delmont, 98
Out-migration, 21-22
Overthrust Belt, 21
Oyster Bar, *102*

P
Park City, 105, *106-107;* arts festival, 105
Parks, *10-11, 23, 108*
Parkview Plaza, 48
Peery Hotel, 102
Phoresor, 60
Pioneer Day Holiday, 15, 16, 21
Pioneer Memorial Theatre, 100
Pollution control, 121
Polygamy, 18
Population, 12, 19, 21; birthrate, 32-33,
    74; county, 28; fertility rate, 74; life
    span, 91; median age, 32; Mormons,
    26; school age, 24, 32, 33, 74; work
    force, 24-25
Prevention programs, 80
Price-Prowswood, 48

Primary Children's Medical Center, 87,
    87
Private schools, *82,* 83-84
ProcureSearch, 45
Project 2000, 26, 73
Promotion of Utah, 50, 52, 53
Psychiatric care, 90
Public transit, 47, 70-71, 72

Q
Questar Corporation, 69

R
Radio stations, 72-73
Railroads, 16, *16,* 17, 67-68
Realms of Inquiry, 84
Recreation, *31, 37, 50, 78,* 110, *112,* 114,
    115, 116. *See also* Skiing; Resorts
Recycling, 121
Redford, Robert, 37, 101
Regional importance, 23-24
Regional/Urban Design Assistance Team
    (R/UDAT), 47, 96-97, 109
Regulation (investments), 49
Religious denominations (non-Mormon),
    18; places of worship, 36-37, 92,
    94-95, 95, 96
Repertory Dance Theater, 98
"Report on the States" (*Inc.* magazine),
    38, 42
Research and development, 54, 56-61
Research Park, 54, 56-61, *56*
Reservoirs, *112,* 117
Resorts, 18, *30,* 34, 49, 50, *51, 52,*
    104-105, *114,* 114-117; traffic prob-
    lems, 117
Restaurants, *22, 102,* 104-105, *104*
Retail, 36, 49, *96*
Ririe Woodbury Dance Company, 98
Robotics, 23
Robson, Thayne, 57
Roper, Roger, 97
Rose, D.N. "Nick," 41
Rose Park, 34
Rowland Hall-St. Mark's School, 83
Rowmark Ski Academy, 83

S
St. Joseph High School, 83
St. Mark's Hospital, 18
Saltair Resort, 18, *30*
Salt Lake Airport Authority, 67
Salt Lake Area Chamber of Commerce,
    18, 41, 42, 79, 81; building, 48
Salt Lake Arts Center, 103
Salt Lake City Municipal Airport II, 65

Salt Lake City School District, 81
Salt Lake City Tomorrow Project, 26
Salt Lake Community College, 86
Salt Lake Convention and Visitors Bu-
    reau, 50, 52
Salt Lake County Mental Health, 90
"Salt Lake County Office Space Studies,"
    46
Salt Lake International Airport, 52, 62,
    64-67, *65;* employment figures, 66; his-
    tory, 64; master plan, 66-67; pas-
    senger traffic, 65, 66; security
    system, 64, 65; snow removal, 67
Salt Lake International Center, 47-48, *67,*
    68
Salt Lake Lutheran High School, 83
Salt Lake-Ogden Metropolitan Statistical
    Area, 24
Salt Lake Police Department, 37
Salt Lake Redevelopment Agency, 47
Salt Lake Tabernacle, *14*
Salt Lake Temple, 12, 16, *27, 36,* 36, 92,
    108
Salt Lake Trappers, 103
Salt Lake Tribune, 18, 73
Salt Palace, *20,* 21, 103, *103,* 104
Salt Palace Convention Center, 53
Sandy City, 34, *34*
San Rafael Swell, 119-120
Santa Fe restaurant, *104,* 105
Save Our Canyons campaign, 116
Scanlan, Lawrence, 92
School closures, 81
School districts, 81
Schools, *76, 82,* 83-84. *See also* High
    schools; Education
Second Century Plan, 21, 46-47
Sensor Technology Center, 57
Services, 60
Settlement, 12, 15-16
Shakespearean Festival, 105
Shopping, 36, 49, *96*
Silver Lake Flat Reservoir, *112*
Silver Lake Lodge, *52*
Silverstein, Joseph, 100
Skiing, *31,* 49, 50, *51,* 83, 105, 110,
    110-111, 112-113, 114, 115, 116; nor-
    dic, 115
Skywest Airlines, 66
Small Business Council, 41
Small businesses, 41-42, 58
Smog, 25
Snowbird, 100-101, *101,* 105, 106, 114,
    114
Solitude, *51,* 114, 115-116
South High School, 81

South Temple Street, *94*
South Towne Center, 36
Space technology, 58
Special education, 79
Sperry, 44
Sports, 103-104, *103*
Squash courts, 107
Stackhouse, Mark, 120
Standard Oil Refinery, *44*
State of Utah Arboretum, 86
Steel, 40-41, 43
Sterling Scholar program, 79
Studio space, 101-102, *102*
Subcutaneous peritoneal access device, 60
Sundance Institute, 101
Sutherland, Ivan E., 59
Swiss Days, 106
Symbion, 60, 61
Symphony Hall, 100, *100*
Synagogue, 36

T
Taxes, 25, 34, 36, 40, 81
Teaching profession, 80
Technovest, 58
Television stations, 72, 73
Temple Square, *14, 36,* 49, *49,* 92, *94*
Terra Tek, 54, 56, *56,* 60, *61*
Theater, 98-100
"This is the Place" monument, 15-16, 21, *21*
Thomas, Richard, 44
Three Lakes Divide, *118-119*
Titanic exploration, 23
Total Artificial Hearts. *See* Artificial hearts
Tourism, 49-50, 52, 92, 96, 110; promotion, 50, 52, 53
Tracy Aviary, 120, *121*
Transcontinental railroad, 16, 17, 67-68
Transportation, 20, 26, 47, 64, 70-72; traffic problems, 116-117
Triad Center, 47, *47*
Trolley Square, 96, *96*
Twain, Mark, 12

U
Uinta Mountains, 113, 118, *118-119*
Unemployment, 18, 20. *See also* Employment
Union Pacific Railroad, 16, 67-68; train, *69*
Union Park Center I, *45*
Union Woods, 48
Unisys, 44

University Hospital, 89
University-Industry Cooperative Research Centers Program, 57
University of Deseret, 12
University of Utah, 12, 22, 23, 54, *54-55,* 56-61, *57, 80-81,* 84, *85,* 89; dance programs, 85-86; Division of Continuing Education, 78, 85; enrollment, 84; facilities, 84, 86; gymnastics, 86, 104; Health Department, 89; library, *85,* 86; Psychology Department, 89; ranking, 85, 86; sports, 86, 104
University Park, 52
University Park Hotel, 54, *56*
Urban blight, 21
Urban planning, 46-47
USX Corporation, 41
Utah Arm, 60
Utah Arts Council, 98
Utah Arts Festival, 47, 103
Utah Business Week, 79
Utah Central Railroad, *16,* 17
Utah Copper Company, 19, *19*
Utah Division of Securities, 48-49
Utah Economic Development Corporation, 26, 41
Utah Endowment for the Humanities, 98
Utah Film Commission, 101
Utah Historical Society, 96
Utah Holiday, 73
Utah Innovation Foundation, 49, 59
Utah Jazz, 103, *103*
Utah Museum of Fine Arts, 86, 103
Utah Museum of Natural History, 86
Utah 100 artificial heart, 61
Utah Opera Company, 79
Utah Pioneer Partnership, 58
Utah Power and Light Company, 70
Utah Shakespearean Festival, 105
Utah Ski Association, 114
Utah Small Business Development Center, 42
Utah Small Business Innovation Program, 58
Utah State Capitol, *24, 31*
Utah Symphony, 79, 98, 100-101, *100, 101*
Utah Technical College at Salt Lake, 86
Utah Technical Finance Corporation, 58
Utah Technology Venture Fund I, 58-59
Utah Transit Authority, 70-71
Utah Venture Capital Conference, 49
Utah War, 16
Utilities, 34, 69-70

V
Valley Bank and Trust, 48
Valley Fair Mall, 36
Venture capital sources, 49, 58-59
Visitors Center, *49*
Vocational programs, 79
Voyager: flight, 44; technology, 23

W
Wallace Associates, 46
Warehousing, 68, 69
Wasatch forest, 110
Wasatch Front, 24, 116
Wasatch Mountain Club, 117
Wasatch Mountains, 28, *28-29, 33*
Wasatch Sports Guide, 73
Water, 25-26, 115, 117-118
Waterford School, 84
Weber State College, 58
Wechslar, Ann, 116
Weekend getaways, 118-120
Wells office building, 48
Wenger, Frederick L., 37
Western Airlines, 65-68
Westin Hotel Utah, 53
Westminster College, 86
White, Raymond, 88
Wilderness areas, *7,* 115, *117*
Wildlife, 62, 64, *64-65,* 115
Wilson, Ted, 37, 53, 86-87, 108, 117
Woodlands office complex, 48, *48*
Woodward Field, 65
Work force, 42, *43. See also* Employment; Unemployment
Work Projects Administration, 19; workers, *19*
World War I, 20
World War II, 20-21, 65
WXIS, 72

Y
Yellowstone National Park, 120
YMCA Hall, *18*
Young, Brigham, 14, 15, *15,* 16, *16,* 17, 97; home of, *95*
Young, Howard, 46
Young Entrepreneur Program, 79

Z
ZCMI Center, 36, *46,* 47
Zion National Park, 119
Zions Bancorporation, 48
Zion's Cooperative Mercantile Institution, 16